PRAISE FOR *COAL*

"Freese makes her points convincingly and eloquently. . . . Freese paints a fascinatingly wide swath." —*Philadelphia Inquirer*

"A thoroughly absorbing history." —*Boston Herald*

"An engrossing account of the comparatively cheap, usually dirty fuel that supported the Industrial Revolution, inspired the building of canals and railroads to move it, and once made London and Pittsburgh famous for their air." —*New York Times*

"Freese's passion for coal is born out of her work. . . . Freese's book is as much about the growing scientific evidence of the damage coal causes to the environment as it is about the social history of the Industrial Revolution." —*Financial Times*

"Freese has a deft style and a knack for explanatory metaphors. And she enlivens her meticulously researched history with anecdotes and surprising facts. . . . Above all, Freese is a strong storyteller who captivates with detail." —*Minneapolis Star-Tribune*

"A rich social, environmental, and political history that ends on a note of warning about the continued use of coal despite detrimental effects on the environment." —*Oregonian*

"Ms. Freese writes her story well. . . . This, then, is a history of coal, an unglamorous substance that Ms. Freese makes glow like its namesake's embers." —*Richmond Times-Dispatch*

"A masterful piece of research and writing by . . . a talented author."
　—*Roanoke Times*

"Fascinating. . . . It lingers hauntingly in the mind."
　—*New Statesman* (UK)

"An engaging and interesting book, tightly documented and consistently readable. Freese makes a passionate plea for a more considered way of treating the earth, its resources and its inhabitants."
—*Daily Telegraph* (UK)

"Eloquent . . . unsparing. . . . The relation between carbon and climate change has seldom been so clearly and readably explained."
—*Scotsman* (UK)

"Freese's fascinating account of King Coal's place in the western world's industrial expansion reminds us of its historical significance and the lengths to which we went in order to mine it." —*Sunday Times*

"Concise and highly readable. . . . Freese has an excellent eye for unusual stories that fix the details of history in the reader's memory."
—*Globe and Mail*

"Engrossing." —*Mail on Sunday*

"*Coal* is required reading for anyone hoping to understand the context of our post-Kyoto environmental dilemma and the very foundation of modern civilization." —*Calgary Herald*

"Absorbing. . . . Freese makes a convincing case that coal represents a grave environmental threat." —*Edmonton Journal*

"Hugely entertaining, and fun to read. . . . There is so much here—the excitement of coal power, the fear of coal pollution, the search for alternative energy, and the possible futures of this planet."
—*Mendicino Public Radio*

"The history of coal, that unglamorous substance that environmental attorney Freese manages to buff until it shines like its distant cousin the diamond. . . . It's dirty, it's cheap, and its past—in Freese's hands—makes for an intriguing, cautionary tale."
—*Kirkus Reviews*, starred review

"Deleterious to health and beneficial to wealth, coal contains a tension that makes its story a compelling one. . . . Freese's combination of labor and technological history is fluid and evenhanded." —*Booklist*

"A strong plea . . . asking governments to remove risk from the act of breathing." —BookPage

"As this human history of coal makes clear, there are no easy answers. But books as lucid as Freese's make a welcome contribution to the search for a sustainable energy economy." —*Natural History*

"[Freese] humanizes her narrative with rich detail and clear, engaging prose." —*onEarth*

"Given the particular chemistry of global warming, it's possible that the decisions we make about coal in the next two decades may prove to be more important than any decisions we've ever made as a species. This book—full of lore, and also of insight—will give you all the background you need to understand why this subject is so vital."
—Bill McKibben, author of *Eaarth: Making a Life on a Tough New Planet*

"In *Coal*, Barbara Freese takes us on an enthralling journey, across time and across continents, using the fascination with coal and the crucial need for it as a way of approaching some of the most fundamental questions of human existence. Her style is engaging, her research impressive, her message an important one."
—Howard Zinn, author of *A People's History of the United States*

"With her abundance of illuminating, often startling insights, Barbara Freese shows us how profoundly we are defined by our energy choices. This epic story illustrates the ways coal has redefined the role of workers, changed family structures, altered concepts of public health and private wealth, and crystallized a profound and enduring debate over national values. Coal has generated social movements even as it has consolidated power structures, and today it threatens to destroy the very civilization it helped create. An engaging book with surprises on virtually every page."

—Ross Gelbspan, author of *The Heat Is On*

"Barbara Freese has a nose for the links between things, technology, and culture. This is a book I'd like to put on my class reading list. I can think of no substance that has played so important a role in shaping industrial technology and the relative fortunes of competing economies."

—David Landes, author of *The Wealth and Poverty of Nations*

"Once Barbara Freese has the reader eating out of her hand with telling anecdotes. . . have you ever considered travelling by rail with a bucket of sand in your lap in case your fellow travelers catch fire?—she turns to sterner stuff. You are by then an addict, and you finish the book knowing a great deal more about global warming, Kyoto, and such than you ever thought you would. . . . A marvelous book."

—Liza Picard, author of *Victorian London*

COAL

COAL

A Human History

BARBARA FREESE

BASIC BOOKS

A Member of the Perseus Books Group
New York

Hardcover published in 2003 by Basic Books,
A Member of the Perseus Books Group
Paperback published in 2016 by Basic Books

Books published by Basic Books are available at special discounts for bulk purchases in the United States by corporations, institutions, and other organizations. For more information, please contact the Special Markets Department at the Perseus Books Group, 2300 Chestnut Street, Suite 200, Philadelphia, PA 19103, or call (800) 810-4145, ext. 5000, or e-mail special.markets@perseusbooks.com.

Library of Congress Control Number: 2002114066
ISBN: 978-0-7382-0400-0 (hardcover)
ISBN: 978-0-7867-2911-1 (hardcover e-book)
ISBN: 978-0-465-05793-1 (paperback)
ISBN: 978-0-465-09618-3 (paperback e-book)

10 9 8 7 6 5 4 3 2 1

Contents

one

A Portable Climate

IN THE SUMMER OF 1306, bishops and barons and knights from all around England left their country manors and villages and journeyed to London. They came to participate in that still novel democratic experiment known as Parliament, but once in the city they were distracted from their work by an obnoxious odor. These nobles were used to the usual stenches of medieval towns—the animal dung, the unsewered waste, and the rotting garbage lining the streets. What disgusted them about London was something new in the air: the unfamiliar and acrid smell of burning coal. Recently, blacksmiths and other artisans had begun burning these sooty black rocks for fuel instead of wood, filling the city streets with pungent smoke. The nobles soon led popular demonstrations against the new fuel, and King Edward I promptly banned its use. The ban was largely ignored, so new laws were passed to punish first offenders with "great fines and ransoms." Second offenders were to have their furnaces smashed.

Had the coal ban held up in the centuries that followed, human history would have been radically different. As it happened, though, in the late 1500s the English faced an energy crisis when their population rose and their forests dwindled. They learned to tolerate what had been intolerable, becoming the first western nation to mine and burn coal on a large scale. In so doing, they filled London and other English cities with some of the nastiest urban air the world had yet seen. They also went on to spark a coal-fired industrial revolution that would transform the planet. The industrial age emerged literally in a haze of coal smoke, and in that smoke we can read much of the history of the modern world. And because coal's impact is far from over, we can also catch a disturbing glimpse of our future.

COAL IS A COMMODITY utterly lacking in glamour. It is dirty, old-fashioned, domestic, and cheap. Coal suffers particularly when compared to its more dazzling and worldly cousin, oil, which conjures up dramatic images of risk takers, jet-setters, and international conspiracies. Oil has always given us fabulously wealthy celebrities to love or hate, from the Rockefellers to the sheiks of the Middle East. "Striking oil" has become a metaphor for sudden, fantastic wealth—riches derived not from hard work but from incredible luck.

Coal does not make us think of the rich, but of the poor. It evokes bleak images of soot-covered coal miners trudging from the mines, supporting their desperately poor families in grim little company towns. Long past the time when it was actually part

of our daily lives, coal is still considered mundane. Earlier generations' familiarity with coal bred contempt for it; and though
the familiarity has faded, the contempt lingers. Even today, children may have heard the warning that if they are bad, they will
find nothing but a lump of coal in their Christmas stockings.
They may never have seen coal, may not even know what it is,
but they know that a lump of it (indeed, a *lump* of anything) is
not something they want. Where oil is seen as a symbol of luck,
coal is seen as a symbol of disappointment.

It's easy, though, to imagine another culture—one with a
greater appreciation of the past, and particularly of the ancient
past—where coal's reputation would be quite different. In that
culture, the lowly lump of coal would be revered as the fossil that
it is. Before mammals appeared, before the dinosaurs evolved,
before the continents glided and crashed into their current positions, that lump was alive. It was part of an enormous swampy
forest of bizarre trees and gigantic ferns—"monsters of the vegetable world," as one nineteenth-century writer described
them—that are no longer found on earth except for some that
survive in greatly shrunken form. Most coal beds were part of the
first great wave of plant life to leave the oceans and colonize the
land, paving the way for animals to do the same and sheltering
them as they took important evolutionary steps. In other words,
coal is the highly concentrated vestige of extinct life forms that
once dominated the planet, life forms that were themselves a
critical link in the chain of environmental changes that made the
emergence of advanced life possible. If coal were not so plenti-

ful, one could imagine it lovingly displayed in museums, placed next to the (generally much younger) dinosaur bones, rather than being burned by the trainload.

Even more fascinating than what went into coal, though, is what has come out of it: enough energy to change the world profoundly. For billions of years, almost every life form on earth depended for its existence on energy fresh from the sun, on the "solar income" arriving daily from outer space or temporarily stored in living things. Like living solar collectors handily dispersed all over the planet, plants capture sunshine as it arrives and convert it into chemical energy that animals can eat. And plants don't just convert energy, they store it over time—holding that energy within their cells until they decay, burn, or get eaten (or, in rare but important cases, are buried deep within the planet as a fossil fuel).

Animals eating plants take that stored energy into their bodies, where they not only store it in concentrated form but disperse it through space. A flock of geese, a pod of whales, a herd of caribou—they are all, on some level, mobile battery-packs. They gather solar energy that falls upon one patch of the planet and deliver it to another as they migrate; in this way, they make life possible for their predators even when, for example, the snow is thick and there is not a green leaf in sight. Life on earth is, in short, a vast and sophisticated system for capturing, converting, storing, and moving solar energy, the evolutionary success of each species depending in significant part on how well it taps into that system.

In the animal kingdom, one of the species that can most efficiently turn the calories of its food into useful mechanical energy is our own; humans need about half the calories that, say, a horse needs to exert the same physical energy. Our metabolisms are astonishingly energy-efficient, and that undoubtedly gave us an evolutionary advantage over other species. Perhaps this advantage helped give us the big brains we needed to figure out yet another way to tap into the stream of solar income captured by plants: fire.

By burning plants—especially plants we couldn't eat, like trees—humanity leapt beyond the physical limits imposed by its own gastric and metabolic systems and released far more solar energy than ever before. It was, of course, a momentous step. Fire is one of the distinguishing features of our species. Only people use fire, if by "people" we include the primates that would eventually evolve into people, because we began controlling fire perhaps some half-million years ago, long before Homo sapiens emerged. This new means of controlling energy reduced our vulnerability to the forces of nature, particularly during the long ice ages that repeatedly gripped the earth, and helped make us fully human.

Eventually, people stopped wandering across the land hunting and gathering food and began to grow it instead, a milestone archaeologists generally consider the beginning of civilization. Fire—and the unusually stable climate that has prevailed over the last 10,000 years—made this settled agricultural life possible. Fire let people clear land for crops (using much the same slash-

and-burn methods threatening our rainforests today) and made digestible the cereals they planted. In these more permanent settlements, people eventually learned basic manufacturing skills, like firing pottery, baking bricks, and smelting metals—ways to make products that would last for societies that would last, at least as long as they had fuel.

Many of these early artisans turned to a fuel that would be an important bridge between wood and coal, and is akin to both of them: charcoal. Charcoal is wood that has already been partially burned. For thousands of years, charcoal was made by heaping wood into large piles, or partially burying it, and then burning it in a slow, oxygen-poor smolder that left behind almost pure carbon. The resulting charcoal burned hotter and cleaner than wood, but the process of making it wasted much of the wood's original fuel content, putting an even greater strain on the forests.

As civilizations and nations grew, trees disappeared, depleted by competing demands for fuel, timber, and land for crops. All these needs drew down the same stores of plant-captured solar energy, and those stores invariably ran short. The size of our fires and our meals, our cities and our economies, and ultimately our populations, were all restricted by the limited ability of the plants within our reach to turn the sun's light into a form of energy we could use.

In this world of tight energy constraints, coal offered select societies the power of millions of years of solar income that had been stored away in a solar savings account of unimaginable size. Coal would give them the power to change fundamental aspects of their relationship with nature, including their relationship with the sun, but it would offer that power at a price.

I HAVEN'T ALWAYS VIEWED coal with such fascination. In fact, until recently, I seldom thought about coal at all. Like most people in developed countries, I had no obvious reason to do so. I wasn't mining it or buying it or burning it, and I hardly ever saw it used. As an environmental attorney for the state of Minnesota, I helped regulate some of the state's coal-burning industries, so I was familiar with the many pollutants coal burning puts into the air. Still, I only vaguely understood coal's sweeping impact on the global environment and on society. What really compelled me to look closely at coal was a case that focused my attention on one of the most profound environmental issues of our time: global warming.

Minnesota is a cold state; our winter temperatures are often the most frigid in the United States, outside Alaska. Lows of minus 50 degrees Fahrenheit are not unheard of in some northern counties. At this temperature, a bucket of water thrown into the air freezes before it hits the ground, bananas get so hard that you can pound nails with them (yes, this has been demonstrated), and exposed skin can freeze in mere seconds. This is not a place where the threat of a few degrees of global warming alarms the average shivering citizen, and, because Minnesota is about as far from an ocean as you can be in North America, forecasts of rising sea levels cause even less concern. Even though we didn't necessarily think of ourselves as living on the front lines of global warming (a naive assumption, as it turned out), Minnesota wanted to have some idea of the larger environmental consequences of its energy decisions. So, a few years ago, the state began a legal proceeding that tried to quantify the impact of its electricity use on global warming. Most of Minnesota's electricity, like that of the U.S. as a whole, comes from coal, so this meant trying to figure out what effect the emissions from our coal-burning power plants would have on the earth's climate.

When the proceeding began, few realized what an exquisitely sensitive nerve it would touch. Representatives of the nation's coal industry, including its most colorful and politically extreme wing, intervened in our hearing, helping to make the contentious administrative trial that followed one of the longest in state history. They brought in a phalanx of scientists who testified that Minnesota should ignore what the vast majority of their colleagues around the world were saying about climate change and argued instead that the climate was not changing except in small ways we were all going to enjoy. Minnesota temporarily found itself on the front lines of the larger national battle over climate change.

The industry's aggressive response was fueled by its recognition that climate change threatens its very existence. Climate change is mainly caused by burning fossil fuels—namely, coal, oil, and natural gas—and of these fuels, coal creates the most greenhouse gases for the energy obtained. Today, the United States burns more coal than it ever has, almost all of it to make electricity.

ALTHOUGH MINNESOTA'S decisionmakers flatly rejected the industry's notion that climate change would be limited to climate improvements, adopting instead the widely held consensus that climate change is a grave threat, we don't yet know whether the proceeding will have any effect on state energy policy. The effect of the case on me personally, however, was dramatic. I was left not only deeply concerned about the changing climate but thoroughly intrigued by the lump of carbon at the center of the storm, this often-overlooked fuel that reveals so much about us and the world we've built. The more I dug, the more I could see that a deep, rich vein of coal runs through human history and underlies many of the hardest decisions our world now faces. Following that vein in the intervening years has taken me far afield—from paleobotany to labor issues, from ancient history to modern geopolitics, and from the massive state-of-the-art power plant a few miles from my home to a primitive little coal mine in Inner Mongolia. This book is the result of that journey.

I'M BY NO MEANS the first person moved to write about the enormous impact of this combustible rock. In the late nineteenth and early twentieth centuries, all kinds of people—engineers, plant scientists, businessmen, and theologians—were inspired to write books and articles for the general public in which they waxed poetic about the glories of coal. Even transcendentalist philosophers had something to say on the subject. In the mid-1800s, Ralph Waldo Emerson wrote this about coal:

Every basket is power and civilization. For coal is a portable climate. It carries the heat of the tropics to Labrador and the polar circle; and it is the means of transporting itself whithersoever it is wanted. Watt and Stephenson whispered in the ear of mankind their secret, that a half-ounce of coal will draw two tons a mile, and coal carries coal, by rail and by boat, to make Canada as warm as Calcutta; and with its comfort brings its industrial power.

This quote stands out not just for Emerson's eerily apt choice of metaphor, but because it captures the world-changing essence of coal. It also reveals the nineteenth century's appreciation of how coal was letting humanity transform nature's cold, cruel world into one more comfortable, more civilized.

Coal was no mere fuel, and no mere article of commerce. It represented humanity's triumph over nature—the foundation of civilization itself. As another writer flatly put it: "With Coal, we have light, strength, power, wealth, and civilization; without Coal we have darkness, weakness, poverty, and barbarism."

The reverence many people felt toward coal was often tinged with defensiveness because they knew that others still looked down on the commonplace fuel. Combating this lingering disdain led to some inspired efforts to elevate coal culturally—to lift it not just to the position of a crucial commodity but to have it recognized as a crucial part of humanity's destiny. In 1850, a popular weekly journal edited by Charles Dickens published a story by an anonymous author that reads like the

"Christmas Carol of Carbon," and reflects this effort most plainly.

In this story, a contemptuous youth named Flashley declares his deep disdain for smutty old coal and those who mine it. He is visited that night by a frightful being rising out of the ashes of his coal fire—a figure black and heavy, with a rough rocky skin and a voice that carries the echoes of a deep mine. This specter takes young Flashley on an emotional journey to the rank and hideous primeval forests where coal had its genesis, and to the terrifying depths of coal mines past and present where miners sacrificed their lives to supply the nation. Finally, Flashley comes to understand the true meaning of coal: that it was divinely placed on Earth so that, in the words of the coal-specter, "man may hereafter live, not merely a savage life, but one civilised and refined, with the sense of a soul within—of God in the world, and over it, and all around it—whereof comes man's hope of a future life beyond his presence here. Thus upward, and thus onward ever."

Coal, in short, would raise up not only our civilization but our very souls. Coal would let us control the external forces of nature, and control our own savage human nature, too. Coal was our species' salvation. Vestiges of this attitude still exist today, though they are more likely to be expressed by the coal industry than by fiction writers.

Some took this view of coal a step further: They saw coal not just as evidence of God's desire to elevate humanity but of His longstanding plan to have Anglo-Saxon Protestants do the elevating. Nineteenth-century British and American observers couldn't help but notice that God had given most of the world's coal to them, or so it seemed at the time. This particular distribution was obviously not just blind chance, an American wrote in an 1856 edition of the *Christian Review,* but a prophecy written long ago in solid rock by God Himself. "A race of men energetic and enterprising; fitted by their natural characteristics, by their mental and moral culture, and by their hold on the pure gospel of Jesus Christ, to be leaders in the onward march of humanity, have had thrust into their hands, unlooked for and unexpected, a treasure, which, if used aright, must secure to them a controlling influence on the affairs of the world."

The belief in coal as a divine and civilizing force may help explain why some missionaries to China paid such close attention to its coal reserves; one missionary touring northern China in the late 1800s wrote in detail about the extent of each region's coal, including its chemical composition. A few decades later, a British visitor predicted that every modern coal shaft sunk in China would "cause China's rising sun to fling out one more shaft of enlightenment to pierce the mists of prejudice." The backwardness of the Chinese people was largely due, he believed, to their failure to release the genie lying "bound and black and mighty beneath their miserable mortal feet."

Of the various metaphors used to describe coal, that of the genie may be the most apt. To see coal purely as a gift from God overlooks the many dangerous strings attached to that gift. Similarly, to see it as just an environmental evil would be to overlook the undeniable good that accompanies that evil. Failing to recognize both sides of coal—the vast power and the exorbitant cost—misses the essential, heartbreaking drama of the story. Like a good genie, coal has granted many of our wishes, enriching most of us in developed nations beyond our wildest preindustrial dreams. But also like a genie, coal has an unpredictable and threatening side. And, although we've always known that, we are just beginning to realize how far-reaching that dark side is.

THIS BOOK BEGINS in Britain, the first nation to be thoroughly transformed by releasing the genie of coal. For centuries, Britain led the world in coal production, and largely as a result, it triggered the industrial revolution, became the most powerful force on the planet, and created an industrial society the likes of which the world had never seen.

We then move to the United States, where coal transformed a virtual wilderness into an industrial superpower with astonishing speed. In the process, coal became the nation's most vilified industry and helped shape many features of the U.S. power structure, both figuratively and literally. Although it is increasingly under environmental attack, America's coal industry today enjoys renewed political power and makes the climate change debate in the United States one of the liveliest in the world.

Finally, we travel to China, where coal has played a surprising role in that nation's long and dramatic history. Today, as China is finally beginning to join the developed world, it has learned that the fuel it is depending on to get it there is threatening the fate of the planet, including China's own particularly vulnerable piece of it.

We can never know where we might be if we had not taken the path paved with coal. We know the world would be altogether different. Probably, we would have urbanized, centralized, industrialized, and mechanized anyway, but decades or even centuries later, on a much smaller scale, and in different places and ways. Without coal, we would have languished longer in the poverty, tedium, and oppression of the preindustrial world, but we might have found a more gradual and humane path out of it than the one we took. Had we taken that different path, we might have less material wealth today, but we might not now be facing the most serious environmental threat we've ever known. Coal has always been both a creative and a destructive force. It is the tension between the two that makes the story of coal so compelling.

The Best Stone in Britain

WHEN THE ROMANS INVADED BRITAIN, among the natural riches they found there were conspicuous outcrops of a velvety deep black mineral. It was declared the "best stone in Britain" by one Roman writer because it could easily be carved and polished into beautiful jewelry. In time, Britain became known for its exports of this prized material, and fashionable citizens back in Rome eagerly adorned themselves with it. Not only were the black trinkets they carved from it stylish, but they had the surprising and mysterious attribute of being flammable as well. They called this mineral *gagate* (a word that over the years changed to "jet," as in "jet black"), which is actually a special form of dense coal. Because they weren't good at telling the difference, though, it seems that many Romans were not wearing true jet but plain old coal—the same stuff that would much later be considered the best stone in Britain for entirely different reasons.

The Romans occupying Britain did more with coal than merely dress up with it; they began burning it, too. Soldiers

burned coal in their forts, blacksmiths burned coal in their furnaces, and priests honored Minerva, the goddess of wisdom, by
burning coal in the perpetual fire at her shrine in Bath. Coal's use
as a fuel was not widespread enough to be directly mentioned by
Roman writers, but traces of it have been found at various Roman
sites in Britain. There's no evidence that anyone in Britain
burned coal before the Romans arrived, with one exception. During the Bronze Age, early inhabitants of southern Wales used coal
to cremate their dead. Perhaps these people saw coal as nothing
more than a convenient way to reduce a body to ashes; but more
likely they invested coal with some mystical properties suitable
for escorting their deceased loved ones to another realm. There
have been few points in history when people could resist attach-
ing some larger meaning to this fuel.

After the Romans left England in the fifth century, the nation
moved into a dark and largely unrecorded period of history. We
can catch a glimmer of these dark ages from an English monk
and scholar known as Saint Bede the Venerable, who in 731 A.D.
wrote a history of England after the Roman era. As it happened,
St. Bede lived in a monastery on the River Tyne in northeast
England, a region blessed with the richest coal deposits in the
country. Just downriver, the town of Newcastle sat on such
ample and accessible coal reserves that it would later become the
most important coal-producing region in the world, and a cliché
for a place saturated with coal.

St. Bede suggests that the Roman practice of burning coal
simply died out after the Romans left, even where the coal could
be plucked most easily from the ground. Writing about the min-

erals of the region, he does mention the great abundance of jet. Like the Romans before him, St. Bede might have believed that all the coal deposits he saw were jet. This chronicler of his time makes no mention of anyone using this mineral as a fuel, but he notes that when the black stone was kindled, its smoke could be used to drive off serpents. So, by the 700s, if coal was burned at all in England, it was apparently not for its heat but for its protective smoke.

THE VAST SUPPLIES OF COAL that lay so undisturbed beneath St. Bede's feet had been waiting there since long before the time of the dinosaurs. Before the Jurassic, before the Triassic, before the Permian, there was the Carboniferous period, a long stage of ancient earth history named after the coal that was then forming. For most of earth's four and a half billion years, its land masses stood utterly barren while the story of life unfolded entirely within the sheltering seas. Only around 425 million years ago did plants begin clinging to the wet shores, and then slowly and tentatively they colonized the continents. By the Carboniferous period—from roughly 360 to 290 million years ago—there was no longer anything tentative about the presence of plants on land. Tangled jungles of remarkable lushness eventually swept across the landscape, hungrily drawing into their cells vast amounts of energy from the sun and carbon from the air.

During the late Carboniferous, the small bit of land that would become Newcastle was very close to the equator; suggested reconstructions of the era show Britain nuzzled up against the then-tropical Greenland and Newfoundland, the ocean formerly between them having closed up. Because tectonic forces had not yet done much lifting, the continents were lower and swampier than today. The region's landscape, like that of most coal forests, would have been dominated by the bizarre *lepidodendron*. These huge trees could grow up to 175 feet tall, and they stood as a vivid demonstration of how far photosynthesis had taken those single-celled ocean dwellers that had accidentally discovered the process a few billion years earlier.

Most of the lepidodendron's length was made up of a straight trunk, up to six feet in diameter at the base, covered with a beautiful lizard-skin bark that gave the tree its name ("lepid" meaning "scale" in Greek). At the top, the lepidodendron branched out into a few short arms bearing narrow leaves up to a yard long. Modern drawings of these plants invariably show these grass-like leaves only at the very top; but some scientists think the leaves might have grown along the entire length of the trunk, which would have given it the look of a massive shaggy green pole. Unlike modern trees, the lepidodendron's interior was soft and pithy. As long as water was plentiful, the internal cells would stay expanded and keep the tree erect; without water, these proud giants of the paleoforest would have weakened, sagged, and finally collapsed under their own weight.

In Britain's primal jungles there also thrived a related tree, the *sigillaria*, which might have looked even stranger to the modern eye. Some types of sigillaria had a long trunk forking once near the very top like a two-headed monster, each head crowned with a large spray of strap-like leaves. Other sigillaria were apparently short and stout, their unbranched trunks six feet in diameter at the base but only about eighteen feet tall. This tree was described in one old paleobotany text (a genre not known for colorful writing) as "grotesque," and more "like a huge barrel than a full-fledged tree." Ancient relatives of the modern horsetail also populated Britain's jungles, probably reaching over sixty feet. And then there were the ferns, primitive and highly successful plants, botanically related to the ones you may have growing in a pot in your home—except they had trunks and were thirty feet high.

This ancient jungle also teemed with bugs, and, lacking much competition on land, many grew to mammoth proportions. Some cockroaches reached a foot in length, the dragonflies had wing-spans of up to thirty inches, and the millipedes reached six feet in length—"as long as a cow," as David Attenborough has put it. Although these invertebrate specimens were certainly impressive, the real evolutionary headlines were being made by another line of creatures evolving on the forest floor: our back-boned ancestors, the amphibians. At the beginning of the Carboniferous, tiny newt-sized creatures not far evolved from fish slithered through the wet undergrowth; by the end of it, massive, fifteen-foot-long monsters were dragging their bellies through the primeval mud. We know this because they left both footprints and belly-prints. It was also during this lush time among the Carboniferous trees that some amphibians took the momentous step of forming hard-shelled eggs, thereby evolving into reptiles, from which would later evolve dinosaurs, birds, and mammals.

Many of the plants of the Carboniferous ended up as coal because they failed to decay the way plants usually do. Normally, when a plant dies, oxygen penetrates its cells and decomposes it (mainly into carbon dioxide and water). As the dense mass of Carboniferous plants died, though, they often fell into oxygen-poor water or mud, or were covered by other dead plants or sediments. Sometimes this burial process was part of a very slow minuet between the coastal forests and the seas. During the Carboniferous, glaciers periodically grew and then shrank in the far Southern Hemisphere, making the oceans rise and fall. As the glaciers melted, the rising seas would step forward and engulf the tropical forests in water and sediment; as the glaciers reformed, the seas would fall back and let the forests step forward again.

Because oxygen could not reach the buried plants and do its disassembly work, the plants only partly decayed, leaving behind black carbon. The spongy mass of carbon-rich plant material first became peat. After being squeezed and slow-cooked by the tremendous pressure and heat of geological forces, the peat eventually hardened into coal. And, of course, it wasn't just the forest's carbon that ended up trapped in the coal, but the energy it had accumulated from the sun over millions of years. Instead of dissipating with the plants' decay, that energy was tucked away into the dark recesses of the earth, at least until the amphibians crawling across the forest floor evolved into creatures capable of digging it up.

THE ENGLISH WOULD IGNORE their abundant coal reserves for four centuries after the time of St. Bede. In the late 1100s, historians finally find references to coal as a fuel. The English didn't call it "coal," though, since that was the name they used for charcoal, a fuel that had by then been used for many centuries. What we call "coal," the English knew as "sea coal," a surprising label for such a deeply terrestrial product, and one that stuck until the 1600s. Why they called it *sea coal* is disputed. Some think it's because the North Sea actually carved coal from exposed outcrops and yielded it up onto the beaches with the sand, where it was first gathered and used by the locals. The more common explanation, though, is that since most mined coal had to be shipped by sea to distant markets, it became inextricably linked to water in the minds of those who burned it.

It's hardly surprising that the coal trade began in earnest along the River Tyne. Its hilly banks exposed portions of coal seams from the North Sea to twenty miles inland. Coal was found and mined in many parts of England during the 1200s, but the coal fields around Newcastle were by far the most important. The seams there were good and thick and, importantly, above the water line. This meant that the mines would stay dry, or could be kept dry with simple drainage tunnels. It also meant that the heavy coal could fairly easily be moved downhill to the river, where waiting boats could float it down the Tyne to the markets of eastern England and, in particular, to London.

Moving the coal from where it lies to where it's needed is a problem that has plagued the coal trade from the beginning. Until the advent of railways, coal was either moved by water or not moved much at all. Once you got the coal to the water, though, the world opened up; shipping it the three hundred miles or so from Newcastle to London cost about the same as carrying it three or four miles overland. In this respect, conditions near Newcastle and in England generally were ideal. The navigable River Tyne lay right at the base of coal-laden hills, flowing, according to one eighteenth-century description, "with solemn majesty as if conscious of the wealth which loads its bosom." Prospective coal customers could easily be reached by water too, since most of the population of this small island nation lived near the sea or along its many rivers. In other words, the link between "sea coal" and water was by no means superficial: The miner working to keep the mines drained may have seen water as the enemy, but England's broad use of coal was possible only because the nation had plenty of water to float the coal to market.

As FATE WOULD HAVE IT, most of the coal along the Tyne was held by what was then the most powerful institution in the world, the Roman Catholic Church. The church controlled a great deal of property all over England, and an even greater share of its coal-bearing lands. Around Newcastle, the church owned most of the seams that would someday provide nearly half of Great Britain's entire output. The actual digging and hauling was done by the serfs of the ecclesiastic estates. In these feudal times, whether it was an aristocrat or a church official holding the property, an estate was still supported by the labor of its serfs. So it was that the first English coal miners would have been virtual slaves, digging coal when they were not plowing fields; the proprietors of what would become the world's most important coal mines were bishops and priors, monks and nuns, who lived along the River Tyne.

Of course, neither the serfs nor the bishops knew exactly what they were digging up. Since coal was being found so close to the surface, many considered it a form of living vegetation. Some even suggested that applying manure to the coal would help it grow.

Legally, estate holders were free to dig up whatever coal was located on their land without concern that the crown would lay claim to it. This was not true across the channel, where European monarchs often claimed minerals found on private land. In England though, under the Forest Charter, signed in 1217 just after the Magna Carta, the crown had already yielded to estate holders the ownership of whatever wood or peat was on their land, and coal fell into that same category. Although the crown still claimed ownership of precious metals found on any land in the realm, this humble and possibly manure-smeared fuel was beneath its notice.

Before long, Newcastle's coal trade was inspiring a class struggle between the church and a group of would-be middlemen. Town-based merchants, mainly former serfs who had been able to earn enough to buy their freedom, tried to seize control of the coal emerging from the nearby mines—owned by the bishop of Durham and the prior of Tynemouth—so they could skim off a share of the profits. The first known act of violence associated with the coal trade was a clash between the merchants and one of these ecclesiastics.

According to court records from 1268, an armed band of town burgesses led by the mayor went to the property of the prior of Tynemouth, burned down his mills, roughed up his monks, and stole from his wharf a ship "laden with sea coal." The town merchants pled in their defense that if the monks traded coal without going through the merchants, not only would they lose their cut, but the king's tax on coal could not be collected. The merchants won, and the prior was forced to tear down his wharf. Church officials in Newcastle and the rising merchant class would wrestle over control of the coal trade for centuries to come, though usually with less violence.

WHEN THE CHURCH and local merchants battled over coal profits in the 1200s, they were fighting over scraps. The coal trade was insignificant, and it would be until the latter half of the 1500s, simply because coal use was not widespread. One of the main reasons so few people burned it was undeniably its smoke, the smell of which the English found disgusting and unhealthy. When Queen Eleanor was visiting Nottingham in 1257, she fled the town because she could not abide the smell of coal smoke and feared for her health.

The smoke that so troubled Queen Eleanor was probably from coal being burned by blacksmiths, or by lime burners making mortar used to repair Nottingham castle. Coal was not being burned in the homes of Nottingham in the thirteenth century because the smell of coal was so thoroughly despised that it was not used domestically. Indeed, most of the coal burned at the time, found close to the surface, was particularly smoky. While the Norman castles and great manors had chimneys to draw the smoke outside, it would be another couple of centuries before this luxury was found in the small houses of ordinary people. The common person's fire sat on a raised stone hearth in the center of the room, away from the wooden walls. The smoke would simply fill the room until it escaped through gaps in the walls or roof. From the standpoint of indoor air quality, most of the English had not advanced much beyond the conditions endured by *Homo erectus,* who, hundreds of thousands of years earlier, had huddled over wood fires inside large huts made of sticks.

Beginning in 1285, during the reign of Edward I, various commissions were set up in London to address the problem of coal smoke, which complainants said had "infected and corrupted" the air. By the summer of 1306, enough coal was being burned by blacksmiths, brewers, and others who needed substantial fuel that there was something of a general revolt against it leading to the ban of its use. Despite new laws threatening steep fines and the destruction of furnaces, coal burning continued to be a problem. Some sources claim that a violator of the coal ban was hanged, tortured, or decapitated (depending on the source), but there's no solid evidence to show that anyone was ever executed for burning coal, and other scholars consider it unlikely. Although enforcement efforts may have dampened coal burning for a while, within a few years, lime-burners and smiths were once again perfuming the air of London with the acrid and pungent scent of burning coal.

The reason for London's rising coal use was simple: The population of the city, and of Britain generally, had been increasing, and the forests were disappearing as a result. Forests near cities were lost first; they were burned for firewood, cut for timber, or cleared for crops and livestock. As supplies of energy from the living forests became more expensive, the demand for energy from the buried ones rose proportionately. The coal sellers, watching cities expand and forests shrink, may well have assumed that their industry was in for a long period of growth. As it happened, though, this particular energy problem would be solved not by coal but by a population crash caused by one of the greatest catastrophes in human history.

THE BUBONIC PLAGUE arrived in Europe on a trading vessel that put into a Sicilian harbor in October 1347. The Black Death was already devastating the populations of China, India, and the Middle East; in its first European onslaught, lasting until 1351, it would kill roughly one in three Europeans, or some 25 million people. Some scholars speculate that it was particularly deadly because the population had already been weakened by years of unstable climate, bad harvests, and famine. At the time, most saw the plague as evidence of the wrath of God; though after careful consideration, the medical faculty at the prestigious University of Paris determined that it was caused by an unusual alignment of the planets that had occurred on March 20, 1345.

In England, the plague would break out again at least three more times before 1400, and the population would continue to drop due to disease and other causes until around 1500. By that time, only about 3 million people lived in England, about half as many as had lived there before the plague. As abandoned farms reverted to forests large enough to support the depleted population, the coal trade slumped.

Those trying to sell coal could not have been helped by their product's resemblance to one of the most unique and grisly symptoms of the Black Death: the buboes, or black swellings of the lymph nodes. One Welsh witness to the plague described the eruption of the buboes as looking like "broken fragments of brittle sea-coal," and described the pain as "seething like a burning cinder." This is not the only time skin inflammations had reminded people of coals. Already the term *carbuncle* had appeared in English, derived from the Latin term for small coal or charcoal. Much later, another disease characterized by ulcerating nodules would be named after the Greek word for coals: *anthrax*.

These grim associations came on top of medieval society's longstanding belief that foul-smelling air in general had a sinister effect on health, perhaps one reason the English so quickly decided that coal smoke was a threat. The same Welsh writer who likened the bubonic swellings to fragments of coal wrote, in typical medieval imagery, of "death coming into our midst like black smoke." People were especially troubled by the smell of sulfur detectable in the coal smoke because they believed that sulfur—commonly known as brimstone—characterized the atmosphere of the demonic underworld. In short, in the Middle Ages coal had quite an image problem, associated as it was with disease, death, and the devil.

COAL'S SINISTER IMAGE persisted even though the richest coal mines, like much of the land in England, were still owned by the Roman Catholic Church. By 1500, though, the church was starting to have a harder time getting to the coal. The coal that could be easily extracted through quarrying or from shallow mines was dwindling. Getting at the deeper coal meant more ambitious and expensive tunneling, and importantly, building the costly means needed to keep the tunnels free of ground water when the mines pushed below the water table.

Making such a major investment would take an act of faith in the future of coal, but this was not the sort of faith the church officials specialized in, and they resisted. In fact, they generally didn't do the mining themselves; instead, they leased the mines to others willing to manage them. The leases were so short, though, that no tenant had an incentive to make the investments needed to expand the mines. It has been suggested that if the mines had stayed primarily in the church's hands, the industry would never have been able to meet the huge rise in demand that was to come. We will never know, though, because in 1527 King Henry VIII decided to end his marriage to Catherine of Aragon, whom he blamed for failing to produce a male heir; this decision changed much in England, including the coal industry.

Upon the pope's refusal to grant Henry an annulment, Henry made his famous break with Rome. Among the many consequences of this momentous step was that it led to one of the greatest property shifts in English history. The church owned at that time perhaps a fifth of the nation's land and wealth, and it had an income nearly three times that of the crown—a dangerous position to be in when the king is desperate for money and when anticlerical feelings are running high. About half of the church's wealth was in the hands of the eight hundred or so religious houses, and before long Henry simply took it away; between 1536 and 1539 he dissolved the nation's monasteries, with the aid of Parliament, and confiscated their property. Many of the richest coal mines in England suddenly became the property of the crown, and through the land sales that followed, the property of the growing class of merchants and gentry eager to find ways to profit from them.

Around Newcastle, the town merchants had essentially won their three-century battle with the church over control of the coal mines. The mines belonging to the prior of Tynemouth, the subject of the 1268 clash between the monks and the merchants, were now owned by the merchants. The bishop of Durham maintained control of his vast holdings a few years longer, but those would eventually be put into the merchants' hands by Henry's daughter, Queen Elizabeth I. The Newcastle merchants, finally in control of the region's coal production, were now in a position to expand the mines aggressively.

WHEN ELIZABETH I ASCENDED to the throne in 1558, England was still held in some scorn by its larger, more powerful, and more sophisticated neighbors. England's trade relations with Europe, still semicolonial, were based largely on the nation's export of unfinished wool cloth and raw materials. England still had only a small presence on the seas, and lagged far behind most of its neighbors in science, technology, and urbanization. All in all, there was little to suggest that England was moving into what many would look back upon as its golden age, and the beginning of its rise to preeminence in world affairs.

England's population and economy, like those of most of Europe, were still on the rise from the lows that followed the plague. The life of the English peasant was improving, albeit slowly; in 1577, one writer notes that the homes of the poor now increasingly held little comforts such as pillows, which replaced the "good round log" they had previously rested their heads on. However, there was a serious threat to that economic growth— the trees on this small island nation were once again disappearing. England's wool industry had become so lucrative that more and more landowners were cutting down the native woodlands to make the lovely green pastures that today seem so natural to the English landscape. Also, the iron industry was gulping down huge amounts of charcoal, using up the forests wherever the ironworks were located.

During Elizabeth's reign, dozens of commissions were sent out by the central government to investigate the wood shortage around the nation, and each one confirmed the serious decline of the forests. Contemporary writers were alarmed about this loss of England's woods, and they wrote of huge forests that had been "greatly decayed and spoiled." This destruction meant not only a fuel shortage, which in itself threatened everyone's domestic comfort and the functioning of nearly every industry; it also meant a shortage of the most important building material of the time. Wood was used to construct just about everything, including homes, furniture, carts, tools, containers, and, of course, ships. The navy considered the wood shortage a national security threat. So laws were passed limiting the taking of wood, and penalties for stealing wood became more severe. In rural Essex, those caught "hedgestealing" were to "be whipped till they bleed well."

The fuel shortage was felt most keenly in the cities, particularly in London. The population of England as a whole was growing, but London's was growing even faster. Of course, as the city grew and as the nearby counties were deforested, wood had to be hauled in from increasingly distant locations. The wood fuel was mainly used for home heating and cooking, but most industrial processes still depended on wood, too. The London breweries alone, according to one calculation, burned 20,000 wagon loads of wood each year. As the shortage became more severe, the price of wood rose far faster than inflation, and the poor, for whom fuel was already a major expense, were under increasing strain.

This was a particularly hard time for London residents to be unable to heat their homes. Europe had by this time entered into its so-called Little Ice Age, a period that would last through the 1700s. On average, this was the coldest period since the last ice sheets had left the Northern Hemisphere; the region's climate was characterized by longer, harsher winters and the occasional freezing over of the River Thames. During the winter of

1564–1565, Queen Elizabeth is said to have taken a daily stroll on the frozen river. In the winter of 1607–1608, Londoners set up the first ice fair on the Thames. Booths sold food and drink, and people enjoyed entertainments such as dancing and bowling. There were a few more such fairs over the next two hundred years, and they grew increasingly elaborate.

If the fuel shortage of the 1500s had continued to deepen, it would eventually have slowed not just the economic growth but also the population growth of London. Like that of most cities of the time, London's birth rate couldn't keep up with its death rate; this was due in part to the periodic outbreaks of the plague, smallpox, and typhus, to which the crowded urban poor were most vulnerable. The city's growth depended on attracting fresh new residents from the countryside at a pace faster than they were burying the dead in the urban churchyards. It is hard to imagine that flow of eager immigrants continuing despite a sustained fuel shortage that would have choked the economy and made urban life even more difficult than it usually was. Eventually, life in London would have become unbearable, and people would have chosen to stay in the countryside, closer to the forests, where at least they could have afforded to heat their homes and bake their bread. Later, as the forests continued to shrink, the fuel shortage might have slowed the population growth of the entire nation. Demographic studies show that in preindustrial England, tough economic times caused people to marry later, lowering birthrates.

But the energy crisis never got that severe for one reason: coal. Domestic coal use surged in the 1570s, and before the end of Elizabeth's reign, in 1603, coal had become the main source of fuel for the nation, though not without complaint. The rich in London tried to avoid using coal, still despised for its smoke, as long as they could. It was said in 1630 that thirty years earlier "the nice dames of London would not come into any house or room when sea coals were burned, nor willingly eat of the meat that was either sod or roasted with sea coal fire." Within a few years, though, the nice dames and the nice gents had succumbed. By the second decade of the 1600s, coal was widely used in the homes of the rich as well as of the poor.

By 1600, London's population had reached 200,000, nearly twice that of fifty years earlier, and it was still picking up speed. (By 1750, London would be the largest city in Europe.) The size of the city allowed for increasing professional specialization prompting the development of commercial, financial, legal, and educational institutions and the cultural flowering for which the Elizabethan age is known. London traders drew England more deeply into the rest of the world; eventually they dominated the international cloth trade and grew rich from the emerging trade with America. Before long, England had evolved beyond its semicolonial status into a world commercial power.

THE PEOPLE OF LONDON could never have brought coal into their hearths and homes without the spread, some years earlier, of that little luxury formerly enjoyed only by the upper classes, the chimney. Even in modest English homes, chimneys had become common by the mid-1500s. Some lamented this development, because they credited the wood smoke that had filled homes in earlier years with both hardening the timbers and protecting the health of the inhabitants. Nonetheless, chimney construction and use spread, enabling people to switch from wood to coal when wood became scarce. The fireplaces and chimneys had to be made much narrower for coal fires than they had been for wood fires to provide the proper draw of air (an architectural change that would promote the employment of very young children as chimney sweeps). The widespread use of chimneys did more than just improve indoor air quality; it forced the energy in the coal to part ways from the attendant pollution—the warmth was channeled into the home and the smoke was sent away to be suffered by the world at large.

It was not long before Londoners' tolerance of coal smoke was once again tested, as more and more home fires were pouring smoke into the city air. In 1578, it was reported that Elizabeth I was "greatly grieved and annoyed with the taste and smoke of sea-coales." In 1603, Hugh Platt, the son of a wealthy London brewer, tried to help the city out with a book titled *A new, cheape, and delicate Fire of Cole-balles, wherein Seacole is by mixture of other combustible bodies both sweetened and multiplied.* (Platt was already famous at this time for authoring a tract on preserving women's beauty.) In *Cole-balles,* he noted that coal smoke was already damaging the buildings and plants of London, and he does not treat the problem as a particularly new one. His patented technique, based on practices he had witnessed on the Continent, involved making briquettes of coal and soil, which he thought, inexplicably, would make the smoke less problematic.

When Londoners began burning more and more coal during the 1600s, and as the city grew larger, its air quality continued to deteriorate. The problem is described in vivid detail in a book called *Fumifugium* (from the Latin *fumo* 'smoke' and *fugo* 'to chase away'). *Fumifugium* was written in 1661 by the noted English writer and minor government official, John Evelyn. Among Evelyn's many interests (art, architecture, horticulture, and politics) was the air quality of London, which he perceived to be much worse than that of other cities in Europe. Thanks to the coal smoke belching forth from various sources, he observed that "the City of London resembles the face rather of Mount Aetna, the Court of Vulcan, Stromboli, or the Suburbs of Hell, than an Assembly of Rational Creatures, and the Imperial seat of our incomparable Monarch." Timothy Nourse, a writer who published an essay on the subject of London's air in 1700, held the same view; despite the considerable charms and glories of London, the thick coal smoke that filled the air meant that "of all the Cities perhaps in Europe, there is not a more nasty and a more unpleasant Place."

Of course, we don't know how polluted the city's air actually was, but some anecdotal evidence is telling. Evelyn describes how the sun was hardly able to penetrate the coal smoke, and how a traveler could smell it miles from London, long before the city was visible. He observed that the smoke left a "sooty Crust or Furr" upon all that it touched, "corroding the very Iron-bars and hardest Stones with those piercing and acrimonious Spirits which accompany its Sulphure." Nourse was also alarmed by the damage the smoke did to buildings; indeed, it left the oldest buildings "peel'd and fley'd as I may say to the very Bones by this hellish and subterraneous Fume."

The material damage was not confined to the outdoors. Evelyn reported that soot penetrated every room, "insinuating itself into our very secret Cabinets, and most precious Repositories," and leaving "black and smutty Atomes" upon everything. Furniture, bedding, and particularly wall-hangings were greatly damaged by the smoke. Because tapestries were destroyed in a few years, "losing their Beauty, and stinking richly into the Bargain," wrote Nourse, wainscoting came into fashion to line the walls instead.

Clothing, too, suffered from coal smoke and soot, requiring frequent cleaning. This not only compounded the already considerable hygiene problems of the city but further increased the distinction between the rich and the not-so-rich. Nourse was particularly concerned over the plight of people of rank but not of fortune because they were going broke trying to wash away the smells and stains of the polluted city air. "In a word, 'tis impossible for any Man to live sweet and clean, to appear polite and well-adjusted amidst so many inevitable inconveniencies, without a vast Expence, which whilst some of more ample Fortunes may bear with; Others (and they too many) of straiter Circumstances, no less ambitious to make a Figure in the World, according to their Birth and Quality, fall into Ruine by living beyond themselves, that they may live in the Company of those of their own Degree and Rank."

It was particularly a problem to be caught in the rain, which washed the soot out of the air and left black spots on whatever it touched. (It is not surprising that Londoners took to carrying defensive black umbrellas in the 1700s.) The soot was then left in a black layer upon the city's notoriously dirty streets, where it would accumulate until it dried and blew about again. Nourse complained about these clouds of coal dust on the streets, writing that "when Men think to take the sweet Air, they Suck into their Lungs this Sulphurous Stinking Powder, strong enough to provoke Sneezing in one fall'n into an Apoplexy." Later, the soot would be washed into the Thames, where, according to Evelyn, it left a visible coating on the bodies of swimmers, even miles from the city.

The impact of the pollution on plant life was evident early on. One of the reasons Platt promoted his *Cole-balles* recipe in 1603 was to reduce the damage to gardens. In 1661, Evelyn wrote that the smoke killed bees and flowers, and that many kinds of flowers could no longer be grown in London. As for fruit trees, "those few wretched fruits" that grew in the city had a "bitter and ungrateful" taste and never fully ripened. By 1700, a book called *City Gardener* had been written; it listed the types of plants thought hardy enough to survive coal smoke "so that everybody in London or other cities where coal was burnt might delight themselves in the pleasures of gardening."

EVELYN HAD NO DOUBT that the pollution he railed against was doing harm to the health of Londoners. He blamed coal for Londoners' blackened expectorations, and for the incessant "Coughing and Snuffing" and "the Barking and the Spitting" in the churches. When his musical friends came to London from the countryside, they complained that they lost three notes in the range of their voices. Evelyn described how visitors to the city commonly suffered a variety of physical symptoms that cleared up again as soon as they left. Overall, he believed that coughs, consumptions, and other lung ailments "rage more in this one City, than in the whole Earth besides."

Moreover, both Evelyn and Nourse had no doubt that coal smoke was killing people. Evelyn wrote that soot produced consumptions that killed "multitudes," and asserted that almost half of all those who died in London died of certain lung disorders. Nourse was particularly concerned about the effect on babies, writing that "from this stinking and smoaky Air it is probably, that young Infants are hardly to be bred up in London; For their new-born Bodies, like tender Plants, or Blossoms, are soon blasted by the Sulphureous Exhalation."

Just because these writers believed that coal smoke was so deadly does not prove much, of course. They also believed that illnesses were spread by "miasmas," or bad-smelling emanations that spread through the air. This miasmatic theory of disease held sway for centuries; not until Louis Pasteur's experiments in the late 1800s would the germ theory of disease be fully accepted. Although the miasmatic theory might have predisposed people to blame coal smoke for their illnesses, attitudes were actually more complex. The miasmas people most feared at this point were related to putrid plant and animal matter, not smoke, which is why some of the English credited the wood smoke that filled their homes before chimneys with keeping them healthy.

One school of thought held that coal smoke, too, could prevent the influx of the more dangerous, disease-causing miasmas, and particularly the plague. Evelyn went out of his way in *Fumifugium* to dispute the optimistic view that smoke protected against the plague; he pointed out that London suffered from the worst air in Europe as well as from a heavy plague mortality. Evelyn claimed in 1661 that the view that coal smoke prevented infections had lost favor among the College of Physicians. Only four years later, though, as the bubonic plague swept again through London, the College of Physicians published a pamphlet recommending that the infectious air be corrected by burning coal, to which might be added any of a list of more fragrant combustibles such as cedar and spices. (The college also recommended the frequent discharging of guns to purify the air.) To the extent that the smoke may have driven away the infected fleas, it could indeed have played a role in reducing the spread of the plague. On the other hand, by weakening Londoners, the polluted air may have made them more vulnerable to the plague and other infectious diseases.

Even today, estimating the effect of coal smoke on public health is difficult. Estimates are largely based on sophisticated analyses of detailed death and illness statistics. Seventeenth century Londoners were a long way from being able to perform such statistical analyses, and yet they were taking important first steps in that direction. As it happened, just as Evelyn was publishing *Fumifugium* in 1661, a man whom many would later consider the founder of statistics was conducting the first methodical analysis of London's mortality records.

The man was John Graunt, an obscure London draper. London had kept death records since the 1500s, mainly to let the rich know when the plague was raging so that they could flee the city. The records were compiled by "ancient matrons" called Searchers, whose unpleasant job it was to inspect all the corpses of the city, make inquiries, and determine the cause of death. Graunt decided the data could be more useful if it was reduced to tables and analyzed. His resulting observations, published in 1662, so impressed Charles II that he recommended Graunt's election to the prestigious and newly formed Royal Society, adding that "if they found any more such tradesmen, they should be sure to admit them all without further ado."

Some of the Searchers' categorizations are quaint, and others just mysterious. Deaths are listed under causes such as Affrighted, Grief, Itch, Piles, Planet, Rising of the Lights, and an ailment known simply as Mother. Still, Graunt's analysis is informative. It confirms that infant and childhood mortality was appallingly high, and that perhaps a third of London's deaths were children under four or five years old. It also shows the prevalence of lung-related deaths, which was the largest category and amounted to between a fifth and a quarter of all deaths. Given what we know today, we can assume that many of those deaths were greatly hastened by the cloud of smoke in which these people lived.

Graunt's own views on coal smoke were somewhat ambivalent. He saw smoke as the chief reason why London's death rates were higher than those in the countryside, and higher than those in years past. He also wrote that while seasoned residents lived nearly as long in London as elsewhere, "yet new-comers, and Children do not, for the Smoaks, Stinks, and close Air are less healthfull then that of the Country; otherwise why do sickly Persons remove to the Country Air?" On the other hand, like others, he credited coal smoke for holding back the influx of miasmatic, disease-causing airs.

None of Graunt's conclusions would today hold up in court (where, indeed, most such statistical analyses eventually end up when special interest groups challenge regulations based on epidemiological evidence). Still, his work represents a fascinating first step in the slow and groping process of trying to understand the impact of coal burning on ourselves and on our surroundings. It was a foundation on which others could build as they analyzed the effects of the even greater levels of air pollution that were to come with the industrial revolution.

A CLEARER UNDERSTANDING of what coal was doing to Londoners and their city in the seventeenth century probably would not have made much difference in the choices they made as individuals. Coal's pollution may have been killing them slowly, but a lack of heat would have killed them quickly. It's been estimated that a poor family in London had to spend at least a tenth of its meager income on coal, and possibly much more, and this was when coal supplies were steady and prices low. This would buy enough fuel to keep a small fire burning in one room for part of the day during the coldest months—in other words, enough to stay alive but not enough to stay comfortable. To enjoy the same warmth from firewood, they would have had to spend perhaps from two to five times as much, and the rising demand would have sent prices even higher. Living as close to the margin as they did, coal was the obvious choice.

By the mid-1600s, Londoners did not merely welcome coal into their homes, they were desperate to have it. More than once it was feared that the populace was actually on the brink of violent revolt when war cut off the coal supply. During these "fuel famines," as they were known, the formerly moribund gardens of London thrived in the suddenly clean air, to the astonishment of their owners. At the same time, the complaints of the poor were "great and unspeakable," and many of them reportedly died from lack of fuel. When coal came back to the city, Londoners snapped it up, willing to watch the city's gardens wither again as long as they could keep their home fires burning.

three

Launching a Revolution

After controlling fire for a few hundred thousand years, our ancestors had devised many clever ways to use its power to transform natural materials into everything from metals, pottery, bricks, and glass to salt, soap, and ale. But in all those busy and inventive millennia of baking and boiling, melting and smelting, it probably didn't occur to them that their fires held another power, too—the power of *motion*. For them, the grueling work of moving matter from point A to point B could be done only by muscle, water, or wind (unless, of course, you were actually willing to destroy the thing you were trying to move, like a patch of forest or an enemy village, in which case fire was ideal). This reality placed a tight limit on the material work a society could accomplish.

In the 1700s, though, a device came along that finally allowed humankind to escape this age-old constraint by turning fuel into motion. The device was the steam engine, and the fuel was coal. The steam engine would prove to be the pumping heart of the industrial revolution, making it perhaps the most important invention in the creation of the modern world. And, as it happened, coal not only "motivated" the engine but also motivated the engine's inventors, who were looking to solve the novel problems posed by having to delve ever deeper into the earth for what had become a life-sustaining daily necessity.

DANIEL DEFOE, who in addition to writing *Robinson Crusoe* also chronicled his own travels in Britain, had this to say after visiting Newcastle in the early 1700s:

> Whereas when we are in London and see the prodigious fleets of ships which come constantly in with coals for this encreasing city, we are apt to wonder whence they come and that they do not bring the whole country away; so on the contrary, when in this country we see the prodigious heaps, I may say mountains of coals which are dug up at every pitt, and how many of these pitts there are, we are filled with equal wonder to consider where the people live that can consume them.

At the time of Defoe's visit, Newcastle had been single-mindedly devoting itself to producing those prodigious heaps of coal for more than a century. By the early 1600s, the area's traditional farming economy had been crowded out by the nation's growing appetite for coal. As London grew and reached out to the wider world, Newcastle's focus narrowed. A new society and culture unique to the coal age emerged.

The expanding mines were worked by rural immigrants who came flooding to the area and were crowded into the hovels the mine operators threw together to house them. They were not welcomed by their neighbors. In one Star Chamber proceeding in the early 1600s, a resident complained that the miners were "lewd persons, the Scums and dreggs of many [counties], from whence they have bine driven," adding that some were thieves, others "horrible Swearers," others "daillie drunkerds, some having towe or three wyves a peece now liveing, others . . . notorious whoremongers." This may have been a great injustice or exaggeration, but it does reflect a common attitude of the time and the tensions that would have naturally arisen between an older, settled class of small landholders and the poorer class of semitransient laborers newly arrived from other counties and crowded into inadequate housing.

The miners and their families, commonly referred to as a separate race of humans, were increasingly ostracized by society. This was something new. According to one historian, "Coal created a new gulf between classes. The medieval peasants and artisans, whatever their disabilities and trials may have been, were not segregated from their neighbours to anything like the same extent as were the coal miners of the seventeenth century in most colliery districts." Over time, the isolated miners developed somewhat different habits and speech. As social outcasts who faced astonishing dangers in providing an increasingly vital commodity, they also developed a fierce sense of solidarity, similar to that of soldiers in wartime. Eventually, they would come to recognize their power to band together and improve their lot in life. From this recognition would later emerge some of the strongest stirrings of the English, and American, labor movements.

In one respect, the seventeenth-century English coal miner was lucky: At least he wasn't a coal miner in Scotland. There, whole families were bonded for life to a coal mine, reduced to a form of industrial serfdom. Sometimes a worker would enslave himself by entering into an agreement with the coal owner, accepting a small sum of money or a pair of shoes in exchange for bondage. The mine owners might also give the miner a gift at the baptism of a new baby, thereby inducing the miner to raise the child to work the mines. Once the child entered the mines, he or she became bound for life.

Typically, mining was a family affair in Scotland, the men hewing the coal, the women and children hauling it to the surface. As they were in the long-past era of agricultural fiefdoms, families came to be regarded as property, and if a mine was sold, they would be sold with it. If they ran away, they could be subject to "torture in the irons provided only for [coal miners] and witches and notorious malefactors."

Paradoxically, force and coercion were not the only way Scottish mine owners held on to their laborers. Apparently, some of them also earned substantially higher wages than other common laborers of the time, suggesting that the miners retained some bargaining power despite their servile condition. The fact that both the threat of torture and the inducement of money were needed to keep people from fleeing the mines tells us something about the conditions they faced underground.

IT IS HARD TO IMAGINE a workplace more dismal and dangerous than a seventeenth-century coal mine. Dark, damp, cramped, and chilly, the mines had ceilings that could collapse on your head, air that could smother you, poison you, or explode in your face, and water that could rush in and drown you or trap you forever. Coal mining was one of the few occupations in which a person faced a very real risk of death by all four classical elements—earth, air, fire, and water. It was probably the most dangerous profession of a dangerous time, vivid and literal proof of the depths to which a society would sink for fuel. One moralist of an earlier century concluded that the need to send people to work in such horrid places was itself evidence that God was punishing humanity for the original sin.

Toiling underground was made even more hellish by the miner's dread that the inexplicable disasters that plagued them were due to demons and goblins haunting the mines. It could not have helped that the mines were filled with eerie signs of past life—like the perfect imprint of a fern deep below the surface— that defied everything the miners believed about the history of the world.

Among the terrors miners faced were three deadly gases that built up in the mines, all variations on the carbon theme, and all deadly exhalations from the coal itself. The first, called "choke damp" (the German *dampf* or 'damp' meaning fog or vapor) is mainly the familiar greenhouse gas carbon dioxide, which in high concentrations can snuff out a life the same way it snuffs out a fire. Choke damp forms when the carbon trapped in coal oxidizes—when the lepidodendron and other prehistoric swamp plants, held in suspended animation for so long, are finally exposed to air and allowed to continue down the path of decay.

Writers in the 1600s described the suddenness with which this invisible gas would suffocate its victims. In one incident, a band of eight men and one woman entered an area of a mine where choke damp had collected and "fell down dead, as if they had been shott." Another account notes that choke damp smothered its victim so quickly that the miner was "without access to cry but once 'God's mercy.'" Miners would sometimes encounter the gas while being let down the mine shaft on a rope at the beginning of a shift and then fall from the rope to their deaths. When exposure to the choke damp was not immediately fatal, those afflicted were treated by their colleagues with "the ordinary remedy": digging a hole in the earth and laying the victim belly down, mouth in the hole. As a last resort, they filled the victim "full of good ale"; and if that failed, they would "conclude them desperate." Some who did survive were noticed to "have some lightness of Brain thereafter."

The second gas plaguing the miners was "white damp," which is carbon monoxide. A product of incomplete combustion, carbon monoxide appeared mainly after a fire or explosion in the mine, sometimes striking down those trying to rescue fire victims. (Carbon monoxide is the same silent poison that can kill you in your sleep if your furnace malfunctions, and it is also one reason you may be required to have your car inspected, since badly tuned cars emit large amounts of this pollutant.) Although carbon monoxide is odorless, white damp is frequently and somewhat inexplicably described in old texts as having a floral scent. One report describes how "an odor of the most fragrant kind—a faint smell of violets—is diffused through the mine, resembling the scent of sweet flowers; and while the miner is inhaling the balmy gale, he is suddenly struck down and expires in the midst of his fancied enjoyment."

It was the threat of the insidious white damp that would prompt miners to bring a caged canary or mouse into the mines because the animals' greater sensitivity to the poisonous gas could provide an early warning. (Canaries would eventually be deemed to work better than mice, in part because a poisoned canary would topple dramatically off its perch, a reaction much easier to detect in a dark coal mine than the "pinkness in the snout" and "crouching" displayed by a poisoned mouse). In the 1600s, though, this precaution was still centuries away. At that time, dogs were the only animals used to help detect the presence of poison gases, and they were sometimes lowered down a mine shaft suspected of containing choke damp. According to one 1662 report, however, mine operators did not bother to lower a dog down the shaft unless the first person sent down on the rope was killed.

Choke damp and white damp took a regular toll in miners' lives, but they were nothing compared to the carnage caused by the third gas, "fire damp," which appeared more frequently as mines got deeper in the 1600s and 1700s. Fire damp is mostly methane, also formed by decaying vegetable matter. The main component of natural gas, methane is commonly known as marsh gas, and is itself a potent greenhouse gas. Fire damp would seep from the coal seams, or sometimes come hissing out quickly from a fissure in the seam. Lighter than air, fire damp would build up along the mine ceiling. Eventually, a miner, holding or wearing an open flame to light his way through the darkness, would come into contact with the gas.*

If there was only a small accumulation of fire damp, the bearer of the flame might merely be knocked flat and singed. If the build-up was large, the results were catastrophic for everyone in the mine. According to a nineteenth-century description of the phenomenon, a massive explosion "with the noise of the loudest thunder . . . sweeps before it into horrible ruin and destruction the unhappy miners, with the horses, carriages and working implements, and dashes, mangles and buries them in one common ruin amid the rubbish and timbers carried along this fiery desolating tempest." Scores of miners could be killed in such a blast. Sometimes the force of the blast was so strong that, according to witnesses, its victims would be shot a considerable distance out of the mouth of the mine like bullets out of a gun.

In an effort to prevent the fire damp from reaching such deadly concentrations, mines sometimes employed a so-called fireman. His unenviable task was to cover himself with soaked rags and creep on his belly along the mine floor while holding a long stick with a lighted candle on the end. He would then raise

---

*Or, the flame might come in contact with the gas another way; a story circulated about an explosion that resulted when a mine rat grabbed a burning candle in its mouth—rats liked to eat the wax—and ran down a passageway into an area where fire damp had built up.

the stick up to the spot where the fire damp was suspected of collecting, causing it to flame along the ceiling while the fireman pressed his face to the floor until it had passed over him. The use of firemen in England didn't begin until the 1600s, but an earlier reference talks about a similar practice in a Belgian coal mine where, in 1554, a mystified visitor apparently observed a miner burning off a jet of fire damp escaping from a fissure in the mine. The visitor described a miner, fully covered in a special linen garment and holding a staff, entering a passage in a mine: "The miner then draws near to the fire, and frightens it with his staff. The fire then flies away, and contracts itself by little and little; having then expended itself, it collects itself together in a surprising manner, and becoming very small, remains quite still in a corner. But it behooves the man who wears the linen garment to stand over the flame when at rest, always terrifying it with his staff."

Miners and mine owners kept trying to find better ways to ventilate the mines, safer ways to burn off the fire-damp, and even alternative forms of lighting that would not trigger an explosion. In some pits, the miners even experimented with bringing phosphorescent fish down to light their way; apparently, this approach didn't fully meet their needs, though it was surely one of the more creative responses to the problem. Despite the various efforts, the fire damp problem became worse as miners ventured deeper.

In certain regions, catastrophic mine explosions became so commonplace that mine owners asked local newspapers not to report on them. The Newcastle *Journal* announced with stun-

ning candor in 1767 that these fiery catastrophes were more common than ever, but "as we have been requested to take no particular notice of these things, which, in fact, could have very little
good tendency, we drop the further mentioning of it." Newspapers avoided disturbing their readership by mentioning the
ongoing deadly explosions for the next several decades. The
public, already complacent over the hardships faced by coal
miners, now had even less reason to think about them.

In addition to the constant threat of gases, miners were also
threatened by various sources of water. Rainwater seeping down
from the surface accumulated in the tunnels, and once the mines
moved below the water table, the surrounding ground water and
the occasional underground stream or spring also contributed
to the problem. In time, these sources could slowly submerge
the mine.

Other water sources posed a more sudden risk. In one spectacular incident, two gentlemen fishing in a river in Scotland
were perplexed to see "a slight eruption" in the water, followed
by a gurgling. When they surmised the river had broken through
to a coal mine beneath, they ran to warn the miners and were just
in time to save their lives. The hole on the surface quickly
became a chasm that caused the entire flow of the river to be
sucked into the mine, including a boat, leaving the fish flopping
on the exposed riverbed. Later, as the pressure of the air that was
trapped in the mine became too great, the water "burst through
the surface of the earth in a thousand places, and many acres of
ground were to be seen all at once bubbling up like the boiling
of a cauldron."

More commonly, though, sudden floods occurred when miners penetrated an abandoned mine of an earlier era that had since filled with water. When that happened, the results were usually much more gruesome, drowning the miners immediately or, worse yet, trapping them in the dark mine until they starved to death.

THE UNDERGROUND WATER that was killing so many coal miners was also threatening the public at large. Indeed, the water in the mines was threatening to cut the nation off from the coal on which it had become so dependent.

In 1610, the operator of one of the largest mines in Newcastle reported with dismay to Parliament that because of drainage problems, Newcastle's coal mines—which many thought were inexhaustible—would not last more than twenty-one years. This dire prediction was overly pessimistic, of course; the British found various inventive ways to keep their mines dry and productive in the decades that followed, but it wasn't easy, and the drainage problem was just postponed.

The first and simplest solution to draining mines was to use gravity, and since early mines were located in the hills, workers could dig long, narrow tunnels that channeled the water into a nearby valley. The British were burrowing through the hillsides with these drainage tunnels as early as the 1300s, and by the mid-1600s there were thousands of such tunnels. Digging the tunnels, sometimes called "watergates," was surely one of the most claustrophobic tasks imaginable. To resist cave-ins and save money, the tunnels were kept as narrow as possible, often no more than eighteen inches wide—just enough to let one worker with a pick dig his way through the hillside. Some tunnels were no more than four feet high, forcing that worker to swing his pick while kneeling.

Many tunnels were more than a mile long, some as long as five miles, with shafts rising to the surface here and there to provide air and allow the diggings to be removed. Usually, construction began from the valley, and workers might push upward into the hill for years before reaching the pit. The challenges involved in keeping the aim accurate enough to reach the designated point in the flooded mine makes some of these tunnels astonishing feats of surveying. They were also enormously dangerous to build because, when workers finally pierced through to the flooded pit, the pent-up water would burst forth with deadly force. A 1665 report explains that at the moment of contact the workers had to protect themselves as best they could from "being dashed in pieces" against the sides of the tunnel.

Eventually, the mines pushed too deep into the earth to be drained this way, and people were faced with the heavy task of lifting the water out of the mines. During the reign of Henry VIII, some mine workers hauled the water up in buckets strapped to their backs, although most mines used various devices that greatly increased the effectiveness of human muscles. Among the earliest was a simple windlass, similar to those used at the tops of old-fashioned wells, that lowered a rope with a barrel attached to the end into the mine shaft. A variation on this design, called a "chain of buckets," was simply that: Buckets were attached along a loop of thick iron chain like charms on a bracelet. This device was suspended on a spool over the mine shaft so that it could hang down into the flooded area. Unfortunately, if a bolt broke, one historian explained, the whole heavy contraption "fell to the bottom with a most tremendous crash, and every bucket was splintered into a thousand pieces." In the early 1600s, there was even a primitive form of vacuum pump that used a leather-covered piston fitted inside a pipe, but these were rare.

Large mines often used many different engines and pumps, both above and below ground, working together in multiple stories to raise water higher than any one device alone could. And they were not all powered by muscle. A few lucky mine operators could raise water with water if there was a nearby stream capable of pushing a water wheel. A few others could raise water with wind, and they experimented with Dutch windmill technology. For centuries, though, most of the water was raised by horses, and horse power wasn't cheap. Large mines might need fifty or sixty horses to keep the engines moving night and day, and the cost of feeding the animals and paying workers to handle them was high.

In time, the battle against the incessant seepage of water helped change the face of the industry—and in some places, the face of society—by accelerating the trend toward ever larger mines and workforces. To recover their large investment in drainage, operators needed to produce more coal. This meant extending the mines and hiring more miners. As the mines expanded, so did the water problem, which, in turn, called for more investment in drainage; and so the cycle repeated. By 1700, coal mines were among the very largest of industrial enterprises in Britain, both in workforce and in investment.

Coal production was by 1700 perhaps ten times greater than it had been in 1550. Coal was a critically important domestic fuel in the cities, and, with some notable exceptions, it fueled the nation's industries as well. It's thought that England was at this time getting far more energy from coal than it could possibly have obtained from its woodlands—even if every woodland in the nation had been managed for optimal fuel production and no wood had been used for building or manufacturing. This dependence was without parallel; by 1700, Britain was probably mining five times more coal than the rest of the world combined. But as the nation's dependence on coal rose, its coal supplies sank, eventually beyond the reach of existing drainage methods.

Despite a century of frenetic invention and huge investment, the flooding mines remained one of the nation's most pressing technological challenges. New ideas were constantly emerging as legions of inventors turned their attention to the problem, but most of the new approaches proved unworkable. As one observer complained in 1708, every year more mines were left "unwrought or drowned for want of such noble engines or methods as are talked of or pretended to." Coal awaited its Edison.

ENGLAND HAD BECOME a very good place to be a scientist or an inventor. In 1660, Charles II had founded the Royal Society, an independent body set up to promote the theory and application of science. This eclectic group included many of the era's greatest thinkers with their widely ranging interests. One historian has noted that under the Royal Society's roof one might have met Isaac Newton, Gottfried Leibnitz, Christopher Wren, Samuel Pepys, John Evelyn, and Robert Boyle: "Religion, nationality, profession—all were subordinated to this ultimate aim of technical mastery, subordinated to a degree perhaps never before found in history."

Among the many topics of interest to these learned men was coal, which was still a tremendous mystery to them and everyone else. In the seventeenth century, some people still believed that coal was alive, derived from "special seeds for its reproduction and growth under the ground." The church preferred another theory that had emerged to explain the plant impressions commonly found in the coal mines. Dead plants had settled underground after "the whole Terrestrial Globe was taken all to Pieces and dissolved at the Deluge"—a reference to the great Flood recorded in the Bible. Basing his calculations on the maturity of the fossil plants found in the mines, one scientist went on to pinpoint the time of Noah's Flood as the end of May. Not until the 1800s would it be commonly accepted that the coal itself, not just the plant impressions, was plant-based and ancient. In the meantime, Robert Boyle, one of the pioneers of chemistry and a founder of the Royal Society, encouraged his colleagues to investigate scientifically whether the miners "ever really meet with subterranean deamons."

The Royal Society members may have been confused about coal's origin, but they were certain of one thing: The nation's supply of it was threatened by flooding. So they focused their attention on pumping technology, and on the concept of atmospheric pressure. Scientists had recently come to understand that the atmosphere has weight, and that it rushes in with great force to fill a vacuum. They knew that the force of air pressing against a vacuum could be harnessed, but creating the vacuum was a huge practical problem. Some tried to create a vacuum by using gunpowder. Denis Papin, a French Huguenot living in England, found a better solution: steam. (Papin had earlier demonstrated his knowledge of steam to the society's members by cooking supper for them in his new invention, the world's first pressure cooker, which he called his "new Digester or Engine for softening Bones.")

Sometime between 1690 and 1695, Papin demonstrated to the Royal Society a small device made up of a piston inside a brass cylinder. When he heated water placed in one end of the cylinder, the resulting steam forced the piston up. When he took the flame away, the steam condensed and left behind a vacuum; the piston was then pushed into the vacuum with substantial force by atmospheric pressure. Papin did not believe it was practical to make this piston device large enough to be useful, however, and so began to explore other methods of using steam.

The inventor who ultimately put such a piston to work was not a member of London's renowned scientific elite, but rather a small-town ironmonger by the name of Thomas Newcomen. Curiously, the members of the Royal Society barely acknowledged Newcomen's invention, and those who did tried to belittle its significance. Some historians believe that Newcomen's humble station was the problem. In the words of one, "to the small scientific world of London whose centre was the Royal Society, it seemed inconceivable that such a man could excel the brightest intellects of the day in the power of invention."

We don't know how Newcomen came about his idea. He might have read about Papin's experiment in the Royal Society's news publication, in which case we can draw a line between the open and invigorating spirit of scientific inquiry that existed in England with full royal support and the most important machine of the industrial revolution. But there is no evidence that Newcomen actually learned of Papin's experiment, and some evidence that this highly practical and inventive craftsperson in the hinterlands hit upon the idea of the piston entirely on his own.

In effect, Newcomen made a device that was similar to Papin's piston, but much larger and with a separate boiler in which to make the steam. After the steam was let into the bottom of the cylinder, cold water was added to condense the steam; the resulting vacuum brought the piston down. As the piston moved up and down, it rocked an overhead beam. The other end of the beam moved the rod of a vacuum pump much like those already used in some mines and powered by horses.

The first Newcomen engine known to have been put to use was installed at a coal mine in 1712. If there was a ceremony marking the threshold humanity was crossing, or indeed if anybody was aware of that threshold, history doesn't record it. The engine was known as a "fire engine," and the term "fireman" no longer meant the man who burned away the fire damp, but the man who kept the engine's fire alive. Newcomen's invention was a huge hit with coal mine operators because it allowed them to pump from much lower depths than ever before. And with one engine doing the work of as many as fifty horses, Newcomen's engine was much cheaper than the horse-driven devices it replaced.

Although Newcomen was snubbed by the scientific elite of his time, and is often overlooked in historical accounts, many of those who do remember his work write about it with reverence, calling it "the most wonderful invention which human ingenuity had yet produced," and crediting him with giving "to the whole world . . . the art of converting fuel into useful power for the benefit and convenience of humanity."*

By the 1760s, hundreds of Newcomen's steam engines (which actually used atmospheric pressure rather than steam pressure, and are sometimes called atmospheric engines) were pumping water from coal mines all over England and Scotland. Because ever greater power was sought, a few of the engines were built to fantastic dimensions for the time; some had cylinders over six feet in diameter and ten feet in length.

The engine had a major drawback, though, which meant that this "most wonderful invention" found little welcome outside the coal mines and the occasional metal mine: It was a fuel glutton, needing huge amounts of coal to keep moving. Of course, coal was cheap and plentiful at the mine mouth and the

---

*Newcomen's portrait was never painted, and his grave has been lost, but a "Newcomen Society" is based in England and has branches in some twenty-eight countries. The society studies the history of technology and keeps Newcomen's name alive.

need for the engine was intense, so despite its ruinous appetite, the steam engine found a large and receptive market there. The coal industry in effect sheltered this emerging technology, keeping it alive for decades until somebody came along with an improvement so dramatic that the steam engine was finally allowed to leave the mines and become an all-purpose industrial workhorse.

JAMES WATT, the son of a carpenter, was born in Scotland in 1736. He was the only one of his parents' five children to survive to adulthood, but he was quite frail and he suffered migraines that would keep him confined to his room for weeks at a time. According to a nineteenth-century biographer, Watt's fragile constitution left him with "an almost feminine delicacy and sensitiveness, which made him shrink from the rough play of robust children"; therefore, during his early years he was educated at home, where he spent his time playing with his father's carpentry tools, and, according to legend, closely observing the steam coming from the tea kettle. When he finally went to school as an adolescent, still sickly and unable to join in sports and games, the other children "showered upon him contemptuous epithets," and his schoolmasters considered him slow and backward for his age. Watt thrived in mathematics, though, and studied the art of making mathematical instruments. It was in this capacity that he was asked in his late twenties to fix a small model of a Newcomen engine kept at Glasgow University.

Like the larger version, the model was extremely wasteful, needing lots of coal to keep it going. Watt realized that as steam was injected and then cooled with water, heat was wasted in the constant reheating and cooling of the cylinder. As Watt later told it, he was walking through a park in Glasgow one day when the solution suddenly flashed into his mind: He would install a separate condenser. That is, the cylinder would be attached to another container that was immersed in cold water and thus kept cool. After the cylinder filled with steam, the steam would pass into this cooler container, where it would condense. In this way, the cylinder itself would always stay hot, and ready for the next injection of steam. According to a good friend of Watt's, "this capital improvement flashed upon his mind at once, and filled him with rapture."*

But turning his rapture-inducing idea into a working steam engine would prove to be an ordeal. Watt found a patron in a local industrialist who was having trouble keeping his coal mines dry with his Newcomen engine, and who hired Watt to improve it. Years went by, but lacking mechanics skilled enough to craft the parts he needed, Watt could not build a successful engine. At one point in the midst of all these years of failure, Watt gloomily concluded that "of all things in life, there is nothing more foolish than inventing," and he repeatedly swore to give the whole thing up.

Then Watt found a new partner, the irrepressible Birmingham industrialist Matthew Boulton, and one of the most cele-

---

*The good friend was Dr. Joseph Black of the University of Glasgow, whose discoveries led to the doctrine of latent heat, and whose theories may well have influenced Watt in his thinking. Dr. Black also discovered the gas carbon dioxide.

brated business partnerships in history was launched. Boulton was more than a manufacturer; he was an industrial visionary who had surrounded himself with people who shared his fascination with technology and his faith in how it could transform the world. They formed a sort of industrial salon known as the Lunar Society. Its members included Joseph Priestley, the scientist who discovered oxygen, sulfur dioxide, and carbon monoxide; Josiah Wedgwood, the pottery manufacturer; and Erasmus Darwin, inventor, poet, and grandfather of Charles Darwin.

Boulton had built his own empire not upon the powerhouse industries of coal, iron, or textiles, but on buckles. He had turned his father's modest buckle factory into a modern hardware "manufactory" that employed up to a thousand workers. Boulton's manufactory was called Soho, and there his skilled artisans made an odd assortment of precision crafted goods, ranging from fancy decorative items such as buckles, buttons, and filagree, to serious scientific instruments like clocks, thermometers and telescopes. Soho became a symbol of modern, high-quality British manufacturing, and it actually drew tourists; Catherine the Great was among the many who came to witness its operations.

Boulton had a longstanding interest in steam engines, in part because the water wheels that powered Soho were unreliable. He had earlier performed his own unsuccessful experiments with the steam engine, and he even corresponded with his friend Benjamin Franklin on engine design. Boulton knew of Watt's work in Scotland, and decided to back him. Watt moved to Birmingham and began working on his engine at Soho in 1773. With the skilled workers, financial backing, and emotional encouragement Boulton provided, Watt was able to build a working engine. The all-important cylinder had been crafted by the pioneering iron founder, John Wilkinson. In 1776, as the revolutionary events in the American colonies were unfolding, James Watt put his first two revolutionary engines to work, one to pump water in a coal mine and one to blow the bellows in Wilkinson's iron foundry.

Although Watt improved the steam engine in myriad ways, his greatest achievement was in getting Newcomen's device to squeeze four times more motive force out of a lump of coal. The steam engine's new fuel efficiency allowed it to leave the coal mines and at last find a welcome in the nation's factories. Watt and Boulton marketed their product as an energy-efficiency device and took their royalties as a cut of the estimated coal savings. Watt would become a national legend, and his engine would be widely hailed as a boon to all humanity.

As for Boulton, he never doubted the enormous potential of what he and Watt were doing. When George III asked Boulton what business he was now engaged in, he reportedly said, "I am engaged, your Majesty, in the production of a commodity which is the desire of kings." The king asked what he meant. Boulton replied: "Power, your Majesty."

WITH ITS COAL and its steam engines, Britain had plenty of power, but its ability to take advantage of that power depended on having the iron needed to build the engines and the factories.

This was a major constraint, because at the beginning of the industrial revolution, iron was still essentially a forest product; you couldn't make it without burning vast amounts of wood, which Britain simply didn't have. Since ancient times, iron had been smelted with charcoal, which would be mixed in direct contact with the iron ore and then fired. Charcoal provided not only the necessary heat but also the carbon needed to promote the chemical reduction of the iron (the oxides in the ore would combine with the carbon and be released as carbon dioxide).

Ironworks were notorious for their rapacious consumption of wood fuel, and they were often forced to move to new locations when they had depleted the local supply. The logical solution, of course, was to start smelting iron with coal instead of wood, but there was a problem: Impurities in the coal would contaminate the iron when the coal and iron ore were mixed together. As a result, long after other fuel-intensive industries had turned to coal, the iron industry was still gulping down great tracts of woodland, and domestic iron production was stifled by high fuel costs. By the 1680s, Britain was importing more than half its iron from more luxuriously wooded countries like Sweden, and that dependency continued throughout much of the next century.

The key to making usable iron with coal was to first bake the coal to drive off the volatiles and turn it into "coke," much the way wood is turned into charcoal. It took a century of experimentation with coke before it was successfully used in the early 1700s to make cast iron (iron that is melted and then poured into pre-made casts). But most of the demand was for wrought iron

(iron beaten into shape while malleable). It was not until the mid-1780s that the technology had advanced enough to allow the industry to use coke for all stages of iron production, both cast and wrought. Britain had taken another major step out of the woods.

Iron production came to depend on coal in another way. Before steam, iron furnaces were kept burning by water; that is, the blast of air needed to keep the fire sufficiently hot was provided by waterwheels. The limited force of the water power (and the fact that charcoal would crumble from the weight of the ore if the furnace was too large) kept iron works relatively small, and they often had to shut down in the summer when the streams fell too low. With steam engines to provide a bigger blast and an ample supply of coke to fuel the process, blast furnaces grew much larger, and were able to pour out lots of inexpensive iron all year long. Instead of being dependent on imported iron, Britain became the world's most efficient iron producer in just a few years. The nation finally had the cheap and ample supply of iron it needed to build its industries at home, and its empire abroad.

Now that iron could be made with coal, a critical organic constraint on the growth of the industrial economy had fallen away. The production of coal, steam engines, and iron spiraled upward at a pace that would have made Adam Smith dizzy, and it was accelerated by the mutually reinforcing relationships that existed among the three. Steam increased the demand for both coal and iron, and also made coal and iron easier and cheaper to produce. Cheaper coal and iron made steam engines cheaper to build and to run, which, in turn, attracted more people to steam power, further increasing the demand for coal and iron, and so on.

The nation's coal industry had already grown relatively large by 1700 just by meeting the domestic and non-iron industrial needs of the nation, but it would expand tenfold by 1830, and double again by 1854. The nation raced to open new coal mines to meet the skyrocketing demand: Between 1842 and 1856 the number of coal mines quadrupled. Coal had completely permeated society. It was not only directly present in the bellies of the steam engines, but indirectly present in the engines' iron cylinders and pistons, in the loom's iron frames, in the factories' iron girders, and later in the iron railroads, bridges, and steamships that would define the industrial age.

Steam power did not create the factory system, but it irreversibly changed the scale, the nature, and the location of industrial enterprise. Early factories, mainly cotton mills, were powered by waterwheels, a technology that had been widely used in Britain and elsewhere since ancient times. (So close was the association between factories and water power that factories were still commonly known as "mills" long into the steam age.) As long as the industrialization of Britain was powered by water, though, it would go only so far, and so fast. Waterwheels could be built only where the flow was strong, where the grade was steep, and where there were no barriers, like other waterwheels. Even in a nation as water rich as Britain, good sites became hard to find. Also, the inefficiency of the waterwheel put tight limits on the amount of power that could be drawn out of the water. As a result, water-powered mills were dispersed throughout the countryside, often up in fairly inaccessible hills, and they were relatively small.

With the shift to coal, the pattern was reversed, reflecting the difference in the power source. Coal spawned much larger and ever more mechanized factories because the power available from underground was so much greater than that supplied by a waterwheel. And, because its energy had already been handily condensed over millions of years, coal concentrated the factories and workforces in urban areas instead of dispersing them throughout the countryside. In short, coal allowed the industrialization of Britain to gain a momentum that was nothing short of revolutionary.

IN ITS IMPACT ON HUMANKIND, the industrial revolution has been compared to the discovery of fire, to the invention of the wheel, and to tasting the apple from the Tree of Knowledge. The term *industrial revolution* falls in and out of fashion, but whatever you call it, something transformative took hold in Britain between about 1780 and 1830. From this point on, means of production would grow larger and more mechanized as factories replaced handicraft industries. People would move to the cities in droves and into ever larger workforces. Markets would become increasingly global, and for many, material wealth would reach undreamed of heights.

Britain's embrace of coal was surely not the only impetus; other economic, technological, and agricultural factors were critical, including the influx of cash from Britain's domination of global commerce and the slave trade. Clearly, though, the industrial revolution could never have taken the shape it took without coal because Britain had depleted its firewood stocks long before the age of fuel-consuming machines. The steam engine, at the time such a potent symbol of technological progress and invention, had to be fed with the ancient remains of long-extinct plants. Britain was creating a new and modern world that was firmly rooted in a past so distant that it still could not be imagined.

Although other European nations followed its industrial example, Britain had a good half-century head start in industrialization and would hold its lead for a long time. In 1830, it produced four-fifths of the world's coal. By 1848, it produced more iron than the rest of the world put together. By the time London held the first World's Fair in 1851, Britain was hailed as the workshop of the world, and its markets and its empire reached global scale. Thanks in large part to its coal—and its ability to turn that coal into motion as well as heat—Britain would remain the most powerful nation in the world until the end of the nineteenth century. It would also become the site of the world's first modern industrial society and the mass-produced problems that came with it.

four

Full Steam Ahead

THE WORLD HAD NEVER SEEN a city quite like nineteenth-century Manchester. What not long before had been a quiet town on the banks of two rivers in England was by the early 1800s the hub of coal-fired industrial development stretching for miles across the English countryside. This unprecedented industrial leviathan was built on white fluff and black rock—the cotton shipped over from the slave plantations of the American South and the coal drawn from nearby mines—supporting the countless steam-driven cotton mills looming over the city. In these mills, where mechanization was reaching new heights, the cotton was spun and woven into enough thread and fabric to clothe a fair portion of the world. And the mills were weaving a new social fabric, too, by creating a larger middle class, a richer industrial elite, and an increasingly isolated class of industrial wage-laborers and slum-dwellers.

At the epicenter of the industrial revolution, Manchester was widely seen as a dual symbol of industrial might and misery, both hailed and feared as a portent of things to come. In 1835, not long after he published his classic analysis of American society, Alexis de Toqueville visited Manchester. He described the dual nature of this unprecedented city in these words: "From this foul drain the greatest stream of human industry flows out to fertilize the whole world. From this filthy sewer pure gold flows. Here humanity attains its most complete development and its most brutish; here civilization works its miracles, and civilized man is turned back almost into a savage."

This new industrial society posed a challenge to the traditional view of civilization as humanity's victory over nature, and of savagery as a life dominated by brute natural forces. James Watt himself once reportedly said, "Nature can be conquered if we can but find her weak side," and his engine became humanity's most potent weapon in the attempted conquest, giving humanity unnatural power and unnatural speed. And yet, as Manchester showed, coal and steam helped create a new kind of savage existence not controlled by nature but virtually severed from it. In looking for nature's weak side, we found our own.

MANCHESTER WAS KNOWN GLOBALLY for its cotton goods, but it was coal, not cotton, that gave the city its characteristic look and smell. No one could visit the city, or even see it from a distance, without being struck by the profusion of smokestacks rising above it and the black smoke that billowed forth. According to an official report, there were in Manchester in the 1840s "nearly 500 chimneys discharging masses of the densest smoke," appalling even those accustomed to the foul air of London.

The lives of factory workers in Manchester, and in the other new industrial cities rising up around Britain, were shaped by the burning of coal just as the coal miners' lives were shaped by the digging of it. Coal made the iron that built the machines the workers operated as well as the factories they worked in, and then it provided the power that made the machines and factories run. Coal gas provided the lights the workers toiled under, letting their work day start before dawn and end after dusk. When they left the factory doors, they would walk through a city made of coal-fired bricks, now stained black with the same coal soot that was soiling their skin and clothes. Looking up, they would see a sky darkened by coal smoke; looking down, a ground blackened by coal dust. When they went home, they would eat food cooked over a coal fire and often tainted with a coal flavor, and with each breath, they would inhale some of the densest coal smoke on the planet. In short, their world was constructed, animated, illuminated, colored, scented, flavored, and generally saturated by coal and the fruits of its combustion.

For many, the ability to unleash the power of coal through steam launched an intoxicating era of optimism and excitement. Steam power opened the door to the sons of carpenters, shepherds, and mine workers, men like James Watt—men who could, with inventiveness and hard work and luck, move up in the world and transform it. The engine was also hailed as a boon to humanity as a whole. There was reason to think that, by lifting the yoke of grueling physical labor, the steam engine would help the poor most of all. Arguably, this is what happened in the very long run: The industrial world that sprang up around the steam engine eventually produced the economic wealth many nations enjoy today. More immediately, though, Britain's sudden ability to turn coal power into mechanical power would draw hundreds of thousands of the nation's poor into a profoundly unnatural life—one that might have been somewhat improved materially, but which was impoverished in virtually every other way.

This is not to suggest that unlocking the power of coal *inevitably* led to these various evils; they were also, certainly, the result of social and economic policies that tolerated and exacerbated the suffering. Still, if you wonder what happens when a society, or certain members of it, suddenly find themselves controlling a tremendous new source of power, Britain's history suggests an answer: Those people will push full steam ahead to turn that power into profit and along the way create towns such as nineteenth-century Manchester.

MASSIVENESS WAS THE HALLMARK of the new industrial age. A massive supply of coal fueled increasingly bigger steam engines, factories, workforces, and cities. By the early 1830s, Manchester alone had seven cotton mills with more than a thousand workers each, and another seventy-six cotton mills that numbered their workers in the hundreds. And the industry was still growing: The ample coal supply removed the bottleneck that reliance on waterwheels had produced. Moreover, the steam engine had created a new incentive to build ever larger factories, because the bigger the engine, the more energy it could extract out of each lump of coal. Since the steam engine was an expensive investment, manufacturers would drive as many pieces of machinery with it as possible. These modern machines had reduced the age-old crafts of spinning and weaving to their most basic elements, and they could churn out such enormous quantities of inexpensive textiles that cottage enterprises simply couldn't compete.

The displaced workers from the home-based workshops, along with farmers driven off their land by the agricultural reforms of the time, were then available to staff the urban mills. They were not forming the first industrial community in Britain; the coal miners of Newcastle, working for wages in large, high-capital mines, had beaten them to it by a couple of centuries. Still, the factory workers of Manchester and the other booming industrial cities were forming something new to the world: a large class of people whose lives were shaped, and in many ways reduced, by machines.

Suddenly, the pace of daily work was set not by the energy of the worker or the demands of the season but by the relentless, regular beat of the steam engine. We now take for granted the notion that machines set the pace of manufacturing, but this was seen at the time as something deeply troubling and unnatural. As one observer of the Manchester factories wrote in 1834, "Whilst the engine runs the people must work—men, women and children are yoked together with iron and steam. The animal machine—breakable in the best case, subject to a thousand sources of suffering—is chained fast to the iron machine, which knows no suffering and no weariness."

Each production task was broken into its most basic units, achieving much greater efficiencies but limiting many workers' jobs to the most simple, repetitive, machine-like movements, all synchronized to the rhythm of the steam engine at the factory's heart. For the first generation of factory workers, the shock of being so thoroughly controlled must have been profound. Former farm workers and cottage craftspeople—people who had worked in relative independence and most of whom had never even owned a clock—were subject to perhaps the tightest daily constraints on personal movement and time that had ever been imposed on an entire class of people.

Twilight no longer brought rest because people had by then figured out how to turn coal into yet another form of energy. In the dim ages that had preceded this enlightenment, darkness could be pushed back only by the meager and expensive flame of candles and oil lamps, and night work was difficult. Because most kinds of coal burn with very little flame, for centuries coal had provided heat but next to no light. In the late 1700s, though, gas made by heating coal and collecting the volatiles started being used as a light source. By 1805, coal-gas lights were keeping the factories so bright that observers cheerily compared them to palaces.

Some factories began to operate round the clock, which at least had the effect of shortening overall hours from fourteen- or sixteen-hour shifts down to two twelve-hour shifts. The marvel of what these workers were doing—working all night by light that had traveled to earth millions of years earlier and had been stored in darkness ever since—probably wouldn't have impressed them, even if they had known about it. This is particularly true because many of the factory workers were mere children. With coal power to substitute for adult muscle, and machinery to substitute for adult skill, factory owners found that children were not only adequate for many jobs but cheaper and far easier to discipline.

The booming coal industry was a leader in the brutal treatment of children, and the steam engine just seems to have increased the ways children could be exploited. Although engines made it possible for the mines to press deeper, ventilation problems were increased. The common solution was the use of "traps," a system of doors in the mines that would keep air currents flowing fast enough to prevent the accumulation of deadly gases. The task of operating these doors was given to the smallest of children working under the most nightmarish conditions, and if these young workers didn't perform properly, the safety of the mine was at risk.

A parliamentary commission that in the 1840s finally began paying attention to the scandal of child labor was astonished that there were not more mine accidents given that "in all the coal mines, in all the districts of the United Kingdom, the care of these trap doors is entrusted to children of from five to seven or eight years of age, who for the most part sit, excepting at the moments when persons pass through these doors, for twelve hours consecutively in solitude, silence and darkness."

One eight-year-old girl described her day to the commission this way: "I have to trap without a light, and I'm scared. I go at four and sometimes half-past three in the morning and come out at five and half-past. I never go to sleep. Sometimes I sing when I've light, but not in the dark: I dare not sing then."

The growing demand for coal prompted operators to expand their mines by following narrower coal seams, but because the new tunnels were often too low for horses or adults to pass through, children were used to haul the coal. The commission described the children's plight this way: "Chained, belted, harnessed like dogs in a go-cart, black, saturated with wet, and more than half naked—crawling upon their hands and feet, and dragging their heavy loads behind them—they present an appearance indescribably disgusting and unnatural." The cruel treatment of children in the mines may have made it easier to tolerate their comparatively milder but more widespread exploitation in the coal-fueled factories.

With men, women, and children all working long hours separately in the factories (instead of working long hours together on the farms or in their cottage industries), domestic life was enormously diminished. Contemporaries complained loudly, noting that having their own incomes encouraged young people to move out of the house much sooner. A substantial part of the rise in birth rates during the industrial revolution appears to have come from young people who married and had children earlier than they had before. Having known no other lives, these young people sent their own children into the factories without fully recognizing what was being sacrificed.

The swelling class of industrial workers—uprooted and traditionless—was an alarming sight to the upper classes, who had no doubt they were witnessing the emergence of something altogether new and momentous. One rural magistrate writing in 1808 described the workers as "a fresh race of beings, both in point of manners, employments, and subordination" (just as the emerging class of coal miners had often been described as a separate race of humans nearly two centuries earlier). One observer writing in the 1830s called this new class of workers "but a Hercules in the cradle," brought together into dense masses by the steam engine. It was observed in 1842 that the steam engine and new manufacturing technology had no precedent and "entered on no prepared heritage: they sprang into sudden existence like Minerva from the brain of Jupiter, passing so rapidly through their stage of infancy that they had taken their position in the world and firmly established themselves before there was time to prepare a place for their reception."

But there had been little time to prepare a "reception." In just a generation or two, the fundamental demographics of the nation changed. In 1833, one observer referred to "the conversion of a great people, in little more than the quarter of a century, from agriculturalists to manufacturers." The 1851 census made it official: More Britons lived in the cities than lived outside them. Britain was one of the first nations in the world to achieve this level of urbanization.

The institutions that would later protect the interests of the working class and the public generally were not in place. Manchester, for example, went from small town to teeming slum long before it formed a local government in 1853. Trade unions were illegal from 1800 to 1824, and did not become a permanent factor in industrial relations until much later in the century, even in the coal mines. Working-class men were not given the vote until 1867. Eventually, institutions and social movements would develop to soften the roughest edges of industrialism. In the meantime, a new ideology would emerge to address the plight of the industrial masses.

In 1842, a prosperous cotton mill owner in Germany sent his son to Manchester to learn the family business at a mill the father partially owned. The son was Freidrich Engels, and he was not eager to learn how to run a cotton mill. But he did want to go to Manchester, which he considered the likely starting point of an imminent workers' revolution. While there, Engels led an astonishing double life: bourgeois industrialist by day, revolutionary by night. He explored the city widely, probing its darkest corners. He wrote vividly about the suffering he saw in his book *The Condition of the Working Class in England* in 1844, attracting considerable attention to the workers' misery. Around this time, he also began his fateful relationship with Karl Marx, with whom he would collaborate on *The Communist Manifesto,* published in 1848. The income Engels continued to earn at his father's Manchester mill allowed him to support himself (and Marx) for years.

The coal-fired, steam-driven mass production of Manchester was pushing industrial capitalism further than it had ever gone, even as it fueled capitalism's most extreme ideological reaction. More important, it was mass-producing the industrial masses themselves—a class of people severed from much of what they had formerly valued, including their traditions, communities, domestic comforts, and independence, and whose waking hours were now controlled by the demanding new logic of the machine age.

LIKE MOST FACTORY TOWNS, Manchester grew rapidly, and little value was placed on aesthetics, health, or anything other than industry. No space was set aside for public parks or patches of greenery, and their absence was sorely felt in the tightly packed slums. A doctor testifying before a parliamentary commission commented on the paucity of public gardens and walks in Manchester, stating that "it is scarcely in the power of the factory workman to taste the breath of nature or to look upon its verdure, and this defect is a strong impediment to convalescence from disease, which is usually tedious and difficult in Manchester."

One 1840s government report notes that the density of smoke in Manchester had "risen to an intolerable pitch, and is annually increasing, the air is rendered visibly impure, and no doubt unhealthy, abounding in soot, soiling the clothing and furniture of the inhabitants, and destroying the beauty and fertility of the garden as well as the foliage and verdure of the country." An 1842 account describes "an inky canopy which seemed to embrace and involve the entire place."

Overall, this was a time when the population of England was increasing, for reasons that are still the subject of debate, but national statistics hide what was actually happening in the growing slums. An 1842 government report on the sanitary conditions of the working class—a report that would help launch the public health movement in Britain—stated that "it is an appalling fact that, of all who are born of the labouring classes in Manchester,

more than 57 per cent die before they attain five years of age."
The report makes it dramatically clear that the high death rates
were a function of both poverty and urban surroundings. The
childhood death rates gave the poor of Manchester an average
life expectancy of only seventeen years; the professionals and
gentry of the city could expect thirty-eight years. By contrast, the
rural poor (taking as an example one region where wages were
reported as half those of Manchester) had an average life span of
thirty-eight years (the same as the well-off in Manchester), and
the well-off in the countryside had an average life span of fifty-
two.

Those who did survive had a pronounced lack of health and
vitality. One observer, writing in 1833, recalled "the robust and
well-made" men who worked in the preindustrial cottage industries, and bemoaned the "vast deterioration in personal form
which has been brought about in the manufacturing population,
during the last thirty years"—a deterioration marked by pallid
skin, sunken cheeks, bowed legs, flat feet, curved spines, and a
general air of dejection. Queen Victoria visited the city in 1851,
and although she approved of the orderly behavior of the Manchester crowd, calling it the best she had seen ("nobody moved,
and therefore everybody saw well"), s–he couldn't help but
notice that they were a "painfully unhealthy-looking population." Declining urban health soon became a national security
issue. During the Crimean War, which broke out in 1854, 42 percent of urban recruits were rejected for physical weakness (compared to only 17 percent of rural recruits), and these were young
men who had already been screened by local recruiters.

A malady that stunted and deformed those living in the shadows of the mills, even before they could be employed within them, was rickets. Rickets mainly strikes infants and toddlers, but it can hit later, too, and severe cases cause permanently bowed and stunted legs, a shrunken chest and pelvis, a curved spine, weakened muscles, and an impaired immune system. The chest deformities and immunity problems predispose victims to bronchitis, emphysema, and pneumonia, and in females the contracted pelvises make subsequent childbirth much more dangerous.

Rickets has an unusual cause and an unusual cure. Humans share with the most primitive algae the ability to photosynthesize part of the solar spectrum into a critical nutrient, ours being vitamin D. We can also get vitamin D from our diets, but even today much of the world gets its vitamin D from sunlight, and sunlight was undoubtedly an even more important source of vitamin D in the past. Cut us off from sunshine without giving us another source of vitamin D and, like plants kept too long in the dark, we will begin to wilt. Our bones will literally soften and bend, eventually taking on a consistency more like cartilage than bone.

Babies raised in the new industrial darkness of the 1800s were vulnerable to rickets for many reasons: They were generally malnourished, no one had the time or space to take them strolling outdoors, and, perhaps most important, Manchester's smoke-blocked sun was no more than "a disc without rays," in Tocqueville's words. In the new industrial cities, rickets reached epidemic proportions among urban children, and came to be known elsewhere simply as "the English disease." In some neighborhoods, doctors reported that every child they saw showed signs of rickets. As recently as 1918, a government report found that not less than *half* the general population in Britain's industrial areas suffered from rickets, and called the disease "probably the most potent factor interfering with the efficiency of the race."

A century before the role of the sun in preventing rickets was established, a doctor testifying before a parliamentary commission investigating night work by children in the factories argued that sunlight was critical to children's growth. He pointed out that the deformities common in the industrial towns were absent among Mexicans and Peruvians, who were continuously exposed to light. These concerns were brushed aside by Dr. Andrew Ure, who wrote a lengthy defense of the factory system in his 1835 book *The Philosophy of Manufactures,* and who was certain that the brilliant coal-gas lighting of a cotton mill was more than adequate to meet the developmental needs of the young "factory inmates."

His unquestioning faith in society's ability to use technology and coal to meet its needs independently of nature was widely shared by the new industrial powers. That faith allowed the captains of industry to cut much of the industrial population off from the sunlight of its own day because they were unable to believe that humanity has a biological need for sunlight that runs bone-deep.

A CRITICAL FRONT in Britain's war against nature was its centuries-old battle against mud. It takes some effort to imagine how isolated premodern life actually was, forcing people as it did to live almost entirely off the land and water nearest them. In rainy Britain, nature enforced this isolation by turning roads and tracks to mud, and the nation could never have industrialized if it hadn't found ways to get unmired.

By the time of the industrial revolution, Britain already had a relatively sophisticated transportation network. This was partly because of its accommodating waters, but partly because of its coal. As the heaviest and bulkiest of daily necessities, coal was the nation's cutting edge cargo, the one that kept forcing it to find new ways to move things.

It was noted in the 1600s that "it is the great quantities of Bulksome Commodities that multiplies ships and men," and commodities didn't get more bulksome than coal. And so, as the coal trade grew in the latter half of the 1500s and in the 1600s, so did the nation's shipping fleet. Already in the early 1600s, more ships were used to move coal than everything else combined. England would no doubt eventually have developed its shipping industry without the impetus of the coal trade; but London's growing dependence on coal left it no choice in the matter, and surely accelerated the nation's maritime investment. Once they had built the ships, the ports, the sailing fleet, and the skills required for the coal trade, the English found it much easier to expand their maritime trade to other commodities and other locations. According to one historian, "the coal trade may be regarded, in short, as a magnet which helped to draw Englishmen to seek their profit and their livelihood in ocean commerce."

The expansion of England's private fleet would prove vital not just commercially but militarily, too. Despite being an island nation, England had not always been a maritime power. Henry VIII built its first real Royal Navy, but it was not strong, and in times of trouble, the nation had to commandeer private vessels. Elizabeth I's navy was more powerful than her father's, but even so, it needed the help of dozens of armed merchant ships to defeat the Spanish Armada in 1588. England's coal ships were particularly important for national defense, and by the early 1600s it was axiomatic that the coal trade was the "chief nursery" for English seamen. Although there were more vessels involved in fishing, they were smaller and of less use to the navy. The sturdier coal ships, with their larger crews, were a vital national asset and could be called up quickly when needed. And when called upon, there was no refusing; the coal ships and their thousands of crew members were pressed into service, by force, many times in English history. In fact, in times of war, those involved in the coal trade demanded additional wages because they ran such a high risk of being forced into the navy.

Paradoxically, the coastal coal trade was another reason a navy was so important in the first place. London and the south of England had become dependent on this fragile lifeline to the north, subject to attack by pirates and foreign powers. The navy was frequently dispatched to escort the coal vessels, in convoys, down the English coast. Still, the coastal coal trade was seen not mainly as a vulnerability but as a national asset, and it came to enjoy what one historian called "an almost superstitious reverence" as the source of England's naval strength. There were even those in the 1600s who opposed trying to find inland coal supplies closer to London because it would have choked off the precious coastal trade.

Long after England became a maritime power, land transport across Britain's notoriously muddy and deeply grooved roads was still arduous and risky. Even royalty often traveled on horseback rather than by carriage until the 1600s, because carriages were so easily trapped in the muck. Goods shipped overland, including coal, were frequently sent by pack horse. In the early 1700s, wheeled traffic was so rare in some rural areas that when one coal mine operator finally introduced coal carts, they "struck . . . terror into the poor country people" and panicked their pack horses. The combined effect of horse hooves and wagon wheels digging into the roadways over the generations turned many roads into little more than very deep trenches. In the 1700s, the roads leading to Birmingham were from twelve to fourteen feet deep in places; it was said that a wagon-load of hay could pass along one road unseen from the sides. When James Watt was born in 1736, a trip from his home near Glasgow to London would have taken about three weeks. When the roads were dry enough to be passable, the holes and ruts made them extremely dangerous, and fatal coach and horse accidents were commonplace.

One response was to build waterways where nature had not. In the latter part of the 1700s, "canal mania" hit Britain; canals were dug all over the nation, allowing boats loaded with goods to be pulled by horses walking on the towpaths alongside. The canals were ultimately used for many cargoes, helping pull the nation together culturally and economically, but the need to haul coal inland was often what prompted their construction. The duke of Bridgewater, called the greatest of the canal-builders, stated that every canal "must have coals at the heel of it." The duke was also a major coal owner, and the canal credited with starting the canal era, finished in 1765, was not only built to bring his coal directly to Manchester but was fed by the drainage water from his mines. The need to move coal was also behind the construction of some of the earliest British turnpikes, the more scientifically designed roads that could stand up to Britain's wet weather.

By the 1800s, it was easier than ever to travel and ship goods around Britain, but it was still not easy enough to meet the needs of the burgeoning manufacturing cities, so dependent on moving unprecedented amounts of raw materials into the cities and then heaps of manufactured goods back out. So, in various places, the industrial class decided that instead of just building more old-fashioned canals, down which horses would tow barges, they would instead invest in something modern enough to meet the needs of the new industrial age: rails, down which horses would tow carriages.

Horse-powered railways were popular because the smooth rails allowed a horse to pull vastly more than it could on an ordinary road, and rails could go where canals couldn't. The idea of rails wasn't new: a primitive form of wooden tracks was probably first used in the Middle Ages in the metal mines of Europe, but rails were put to greatest use in the coal mines of Britain, particularly around Newcastle. Before the use of rails, coal was hauled by wagon over muddy trails to the water. But as the coal face was pushed farther from the river, this overland trek became much harder, and in very wet weather the trails became impassable. By the early 1700s, though, horses were commonly pulling the coal carts across wooden rails instead of across the bare earth; miles of these rails, called wagonways, were laid across the Newcastle region. Sometimes the wagonways passed right next to the meager cottages of the coal mine workers. In one such cottage, in 1781, a boy was born who would go on to "free the steam engine and send it thundering across the world on iron rails," and earn himself the title of the "Father of Railways."

GEORGE STEPHENSON'S LIFE STORY is one seemingly designed by Victorian school masters to inspire poor boys, and indeed that is exactly how it would later be used. One of George's first responsibilities was to keep his younger siblings from being crushed by the horse-drawn coal wagons that passed over the rails just two yards from the front door of the one-room cottage where he lived with his parents and five siblings. Stephenson's father was a "fireman" for the local coal mine, meaning he shoveled the coal into a Newcomen steam engine, and this engine fascinated the boy. When he was fourteen, he became the assistant fireman to his father, and his mechanical skills were so phenomenal that by the time he was seventeen, he was put in charge of his own steam engine.

At this point, according to one biographer, Stephenson wanted to learn about the new Watt engine, but first he had to enroll in night school to learn how to read. When he turned twenty, he held one of the best-paying jobs the mines offered, that of operating the winding equipment. When his twelve-hour shift was over, Stephenson made clocks and shoes for a little extra income. He married and fathered a son and a daughter, but then his life took a turn for the worse. His daughter died shortly after birth, and his wife died the next year of consumption. Then his father was scalded and blinded by escaping steam from a steam engine. In his mid-twenties, Stephenson was left supporting his baby son as well as his father, whose debts he had to pay off. When things couldn't look worse, he was drafted to fight in the Napoleanic Wars, and had to give up the rest of his meager savings to pay a substitute to take his place in the militia. He sank into a depression, and even considered moving to the United States, but gave up the idea and kept working at the mine.

His engineering talent earned him great respect, and in 1812 Stephenson was put in charge of the machines used in all the coal mines owned by the Grand Allies, a cartel that controlled much of the nation's coal. The Grand Allies were among the richest people of the early industrial era, and, trusting deeply in Stephenson's skills and inspired by his confidence, they were willing to spend the money needed to carry out his ideas. Stephenson quickly began to look for ways to use steam engines instead of horses to move the coal down the tracks.

George Stephenson is sometimes given credit in English folklore for having invented the locomotive, but this is untrue. The Cornishman Richard Trevithick was the first to use high-pressure steam, rather than the low-pressure steam that powered

Watt's stationary engines. But Trevithick could never pull together all the elements of a successful railway. Stephenson could, building for the Grand Allies the first railway with the sort of flanged wheels and iron tracks that future railways would use.

The difficulty of hauling coal had always been one of its greatest drawbacks as a fuel, but now, through the locomotive, coal could haul itself; similarly, through the steam engine, coal could pump the mines that contained it. The patterns were the same: Coal created a problem, then helped power a solution, and that solution would have revolutionary consequences far beyond the coal industry.

In 1825, Stephenson made history with the opening of a twenty-six-mile railway between the coal town of Darlington and the river town of Stockton. The opening was attended by thousands who watched the procession of thirty-four wagons, carrying not just coal but six hundred passengers as well. It used locomotives on the flat sections of track and moved at a speed so slow that the procession was led by a man on a walking horse; where the track was steep, the cars had to be pulled uphill with ropes and stationary steam engines.

It was 1830, though, when the railway truly burst onto the public scene with the opening of the Liverpool and Manchester Railway, the first fully locomotive-driven public railway, and the one that launched the railway frenzy to come. Stephenson, now fifty, engineered the track, and convinced the owners to use locomotives instead of horses or stationary engines by winning a much-publicized national contest with a locomotive of his own design. Before the railroad's official opening, Stephenson took the fortunate few out for a ride, including the celebrated young actress Fanny Kemble, who was famous in Britain for her recent starring role in *Romeo and Juliet*.

Many of Kemble's contemporaries found the locomotive simply terrifying, "a huge monster in mortal agony, whose entrails are like burning coals." It is hard today to appreciate what a quantum leap in speed and motive power these engines represented, and the awe, even horror they could inspire. A few years later, a country clergymen would tell of taking his clerk to watch a train pass by for the first time. As the engine roared past spewing "dense columns of sulphurous smoke," the clerk "fell prostrate on the bank-side as if he had been smitten by a thunderbolt! When he had recovered his feet, his brain still reeled, his tongue clove to the roof of his mouth, and he stood aghast, unutterable amazement stamped upon his face." Five minutes later when the clerk was able to speak, he asked, "How much longer shall knowledge be allowed to go on increasing?"

To Kemble, by contrast, train travel was a pure thrill. Stephenson pushed the engine to a record thirty-five miles per hour, or, in Kemble's words, "swifter than a bird flies (for they tried the experiment with a snipe)." She had been given the chance to be among the first people to travel at speeds completely beyond the human experience, and was one of the few to write down how it felt. "I stood up, and with my bonnet off 'drank the air before me.' The wind which was strong, or perhaps the force of our own thrusting against it, absolutely weighed my eyelids down. . . . When I closed my eyes this sensation was quite delightful, and strange beyond description; yet strange as it was, I had a perfect sense of security and not the slightest fear." She also declared herself "most horribly in love" with George Stephenson. Had this been a Hollywood movie, they would have lived happily ever after; instead, Kemble moved to the United States and entered into an unhappy marriage with a slaveowner.

The official opening of the Liverpool and Manchester Railway was intended to be dramatic and unforgettable, and it was, but not exactly as the organizers had planned. As many as 400,000 people lined the route on that rainy September day in 1830. The duke of Wellington, who had defeated Napoleon at Waterloo fifteen years before and who was now the prime minister, was the guest of honor. He attended the festivities even though he disliked the concept of the railway and was sure nothing would come of it.

Many other dignitaries rode on the duke's train, including a member of Parliament from Liverpool who was a great champion of the railway, William Huskisson. The duke and Huskisson, both Tories, had suffered a falling out over parliamentary reform policy. Party members hoped that during the opening festivities the duke and Huskisson would reconcile and bring about a coalition that could save the duke's embattled government. A procession of seven trains carrying almost seven hundred invitees began the trip from Liverpool to Manchester.

Midway, the duke's train stopped for water, and Huskisson and some of the other honored guests got out for a stroll. The duke, still on the train, nodded to Huskisson, stretched out his hand, and Huskisson grasped it while standing on a parallel track. Before the important political reconciliation could develop further, though, one of the other trains in the procession came racing down the parallel track. There was much shouting, and various ambassadors, counts, and lords went leaping in all directions. Huskisson was less agile and, bewildered by the frantic scene, he hesitated. According to one witness who recounted the story to Fanny Kemble, Huskisson then "completely lost his head, looked helplessly to the right and left, and was instantaneously prostrated by the fatal machine, which dashed down like a thunderbolt upon him," crushing his leg.

George Stephenson, who had been driving the duke's train, jumped into action. He quickly unhitched the engine and the wagon that had been carrying the full military brass band, had Huskisson loaded onto the bandwagon, and then raced him to a nearby town for first aid. Under other circumstances, Huskisson, an avid supporter of train travel, might have appreciated participating in the setting of a new land speed record—a phenomenal thirty-six miles per hour—with the Father of the Railway himself at the train's controls. Unfortunately, later that night, Huskisson would become the first person to die from a train accident.

Meanwhile, the entourage of dignitaries arrived, several hours late, in Manchester where an impatient crowd waited to greet them. The crowd included many mill workers who were hostile to the duke because of his opposition to reforming Parliament to better reflect the new industrial demographics of the nation. The duke's train was stoned by members of the increasingly unruly crowd, and it had to beat a hasty retreat out of the city. The crowd's behavior no doubt helped confirm the duke's belief, expressed the next year, that the "lower orders" were "rotten to the core," and that revolution was imminent. He strongly opposed changes that would extend the right to vote beyond the upper classes. Exactly two months after the opening of the Liverpool and Manchester Railway, the duke of Wellington's conservative government collapsed, to be replaced by one slightly more open to reforms. Years later, the duke would give up his opposition to rail travel and reportedly make a great deal of money speculating in railway stocks.

The opening of the Liverpool and Manchester Railway was not an auspicious beginning for the railway era, but it did vividly reflect the momentum, excitement, danger, and social turmoil of the early coal-and-steam era. It marked a moment of acceleration in the speed of industrialization, and it fed the growing myth that technological progress was simply unstoppable. In the months to follow, hundreds of thousands of passengers would ride on the Liverpool and Manchester Railway, firmly establishing the future of rail travel and sparking a massive investment in connecting the nation, and then the world, with rails. By 1845, just fifteen years later, Britain had laid down 2,200 miles of track, and that would increase to 6,600 miles by 1852. To many, the railway was a harbinger of a brighter future. Many envisioned that it would finally draw the nations of the world together under the banners of science, harmony, and prosperity. But above all it was, in the words of one nineteenth-century British lord, "a great, a lasting, an almost perennial conquest over the power of nature."

MANCHESTER WAS A CITY largely created by the industrial
revolution, and was thus a compelling symbol of the era, but it
never competed with London for sheer size and power. With
nearly a million residents in 1800, London was ten times larger
than Manchester, the second largest city. Although both cities
were growing fast, London would remain vastly larger; by the
early 1860s, 3 million people would call London home. It was
the largest city in the world, and the center of the world's
strongest empire. Thanks in no small part to the coal-fired
industrial might of Britain, London was widely considered "the
centre of the commercial and political world," and not just by
Londoners.

London's industries were much smaller than those of Manchester and other industrial cities, but it had so many small
industries, and more important, so many domestic fires, that
early in the 1800s it was aptly called "a volcano with a hundred
thousand mouths." The appalling pollution so vividly described
by Evelyn in 1661 had only worsened.

As happy as Britain was to embrace steam technology, it
resisted another form of technology that would have greatly
reduced its pollution problems: the stove. Other advanced
nations had, by this time, widely adopted stoves, which can
warm rooms much more thoroughly and with half or even a
quarter as much fuel as a fireplace, which sends most of its heat
up the chimney.

The British, though, couldn't bring themselves to adopt the
abhorrent devices. They hated to lose sight of the cheery flames
in their hearths, and it's likely that the smokier and darker the

city grew, the more attached they grew to the brightness of their fires. Forced to choose between seeing the sun and seeing their own fires, they chose the latter. (This attachment to open fires would prove so strong that in 1920 one expert estimated that the majority of homes still were warmed by nothing but open fires burning solid fuel, almost exclusively coal.)

Ironically, the British resisted the fire-concealing stoves not just on aesthetic grounds but on air quality grounds. Stoves, it was argued, *burned* the air in a room rather than warmed it. Floating dust particles contacting the hot metal left a singed smell, and some speculated that other mysterious changes to the air rendered it not just smelly but unhealthy. And, by limiting the substantial flow of air up the chimney, stoves reduced the amount of outdoor air drawn into the room, cutting down on ventilation. The British clung, therefore, to their traditional fire-places, burning far more coal than necessary in exchange for the comforting view of their fires, the privilege of avoiding the smell of burnt dust, and the dubious benefit of having a heating system that drew a steady draft of often highly polluted outside air into their homes.

On most days, the air of London was still cleaner than Man-chester's, but under certain cold and windless conditions, the combined effect of smoke and fog could plunge London into almost complete darkness in the daytime, stopping much of the city in its tracks. Of one such day in 1812, the *Times* reported that "for the greater part of the day it was impossible to read or write at a window without artificial light. Persons in the streets could scarcely be seen in the forenoon at two yards distance."

This was unusual enough to make it into the next day's paper, but not strange enough to warrant more than a brief mention.

London had long been known for its fogs, including occasional "Great Stinking Fogs," as one seventeenth-century astronomer called them in his weather records. There was, though, a marked increase in the frequency of London fogs between 1750 and 1890. Natural climate fluctuations could have contributed to the increase, but inevitably so did the smoke. The city could barely be seen from above, such as from the top of St. Paul's Cathedral. Lord Byron described the London skyline as "a wilderness of steeples peeping on tiptoe through their sea-coal canopy." Smoke was such a quintessential part of London that for a long time the city held the nickname "The Big Smoke."

While the fogs were unnatural, they could also be beautiful. Many residents and visitors looked on the smoke as picturesque, enhancing the mystery and excitement of the city, perhaps one reason why the nineteenth-century efforts to clean the city's air were such a failure. British writers called the thick air London's "sublime canopy that shrouds the City of the World," and referred to their "beloved smoke."

Indeed, although some travelers from abroad were appalled by the smoke, many others were charmed by it. A Canadian visitor writing late in the century said the comforting smoke "gave a kind of solidity and nutriment to the air, and made you feel as if your lungs digested it," in contrast to her "monotonously clean" home. In 1883, an American poet living in London wrote, "Today [we] are having a yellow fog, and that always enlivens me, it has such a knack of transfiguring things. . . . Even the cabs are rimmed with a halo, and people across the way have all that possibility of suggestion which piques the fancy so in the figures of fading frescoes. Even the gray, even the black fogs make a new and unexplored world not unpleasing to one who is getting palled with familiar landscapes." For reasons we still don't understand, London's fogs were more frequently being reported as strangely colored, including the occasional orange and dark chocolate fogs.

With all its potential for aesthetic appeal, it's quite clear that the pollution of Victorian London could bring confusion and death, particularly when a dense fog planted itself over the city for three or more days. On a Wednesday in December 1873, for example, a thick, cold fog settled in and stayed until the following Saturday, at one point stretching fifty miles from the city. Visibility was reduced to a few yards, and according to the brief articles printed in the weather section of the *Times*, "all locomotion, especially to people on foot at the crossings of great thoroughfares, became extremely dangerous." The fog was blamed for a sad and chaotic parade of sometimes fatal accidents that left the hospitals straining to care for the injured: Carriages crashed into light posts and other obstacles, people were run over by omnibuses and Hansom cabs (one poor fellow was run over by both), horses tripped and crushed their riders, and a train struck a man who was putting fog lights along the track. The police were kept busy trying to handle the multitudes of lost children. The most common reported fog-related deaths were the twenty drownings that took place over the four-day period, caused when people simply stumbled into the Thames or the canals of the city, or when barge collisions threw them into the water. The paper failed to report that several cattle, in the city for an exhibition, dropped dead in the fog, and many others were in such distress they had to be immediately slaughtered.

Amidst the various locomotion-related deaths, the *Times* also noted the deaths of two gentlemen who in separate incidents fell in the street and shortly died from "inhaling the fog." In reporting just these two deaths by inhalation, the *Times* was off by two orders of magnitude. An analysis of the city's death statistics performed decades later showed that the 1873 fog quietly killed from 270 to 700 Londoners. Another foggy week in 1880 would kill from 700 to 1,100 people, and one in 1892 would kill about 1,000. These deaths went largely unreported, while people worried instead about the far less deadly but more dramatic confusion on the streets.

Nineteenth-century Londoners probably had the statistical skills to detect these deaths long before they actually did, if they had only looked. This was, after all, the city where John Graunt had shown two centuries earlier how much can be learned by counting the dead. Perhaps they didn't look because they had been living for so long in a fog of their own making that they simply took it for granted. They stopped asking the harder questions about the impact this unnatural new world they'd created so energetically was having on its human inhabitants.

This brashness was just one of the traits that, along with coal, put the British at the vanguard of the industrial revolution. Others included the desire to conquer nature, a faith in technology, a belief in private enterprise, tolerance for the misery of the working class, and a conviction that their nation had a larger destiny as a world power. Of course, across the Atlantic there was another nation that shared all these traits and that had even more coal, and it was undergoing an even more astonishing transformation.

five

A Precious Seed

THE FIRST ENGLISH COLONISTS to come to North America were looking for gold and silver and a river route to China. What they found instead were trees, and lots of them—forests of a density not seen in their homeland for thousands of years. The settlers of Jamestown, Virginia, couldn't send home gold, so they filled the ships with timber instead. A second wave of settlers, landing hundreds of miles up the Atlantic coast, also found an abundance of trees. When the Pilgrims emerged from the Mayflower after two months of huddling around coal fires below deck, they stepped onto a land "wooded to the brink of the sea."

To the Puritans, who soon followed, the new continent's trees would become a crucial piece of advertising. Francis Higginson, America's first Puritan reverend, immediately wrote a pamphlet that was published back in London and no doubt helped to attract the thousands of English soon to follow. He wrote: "A poor servant here . . . may afford to give more wood for timber and fire as good as the world yields, than many noblemen in England can afford to do. Here is good living for those that love good fires."

To the masses warming themselves by scanty, smoky coal fires in the crowded and chilly slums of London, the promise of a big crackling wood fire in the spacious new world must have been quite an enticement. Higginson also boasted of the region's "extraordinary clear and dry air," which he credited with clearing up his chronic health problems. He died within months of writing these words, after only a year in the new world. Amidst their new-found wealth of trees, eighty malnourished members of Higginson's flock also perished that first winter, tragically demonstrating that "good living" takes more than just ample fuel and fresh air.

What the settlers in Virginia and Massachusetts didn't know was that they were trying to eke out a life in different parts of the same vast forest, which stretched almost uninterrupted, south to Georgia, north into Canada, and then west, spreading over the mountains all the way to the Great Plains over a thousand miles away. They were perched on the edges of the great eastern American forest, one of the largest stretches of woodland ever to grow on the planet.

To many of the English settlers, the standing forest was a constant source of dread. William Bradford, the Pilgrim governor, described the land the Mayflower had sailed to as "a hideous and desolate wilderness." The woods abounded with "snakes and serpents, of strange colors and huge greatness" (including rattlesnakes). This was not a complete wilderness;

native people burned the woods to reduce underbrush, increase game species, and make clearings, and had probably done so since the forests first spread north after the last ice age. Still, compared to anything the colonists had ever seen, this was a wild, dangerous, and uncontrolled landscape, and the settlers set about controlling it. That meant farming, which meant chopping down the trees to make way for crops and animals, and this they did with gusto.

They built log cabins, log fences, wooden furniture, and tools, and they also shipped timber back to England. Most of the wood went up in smoke, though, burned with abandon in enormous fireplaces (into which they rolled logs too heavy to lift) or simply consumed in bonfires in the newly cleared fields. With more trees than they wanted, the colonists had no reason to search for coal. It would therefore be a very long time before anyone realized that in addition to this astonishing wealth of wood, the continent also held the world's richest coal deposits, including a coal field half the size of Europe lying beneath the eastern American forest.

By the time this carbon treasure was finally discovered, its phenomenal size would be seen as further proof of America's special destiny. Coal would be dug the way forests had been cut—not simply for survival and comfort and profit, but in service of the larger mission of transforming the wilderness into something that transcended nature. As one theologian would put it, the coal deposits had been "scattered by the hand of the Creator with very judicious care, as precious seed, which, though buried long, was destined to spring up at last, and bring forth a glorious harvest."

BY THE MIDDLE OF THE 1700S, the mostly British colonists had firmly established themselves up and down the Atlantic coastal region. They were eager to press farther west, but were being held back by, among other things, the Appalachian Mountains. Once they had crossed this formidable mountain range, the continental interior would open up to them. There the rivers flowed west, and north of Virginia many of them gathered into the magnificent interior river known to the English as the Ohio. In the 1750s, a Philadelphia mapmaker surveyed the Ohio River valley and reported that along with the valley's many riches, "coal is also in abundance and may be picked up in the beds of the streams or from the sides of exposed hills." In the next century, the famed British geologist Sir Charles Lyell would visit the region, about which he wrote, "I was truly astonished . . . at beholding the richness of the seams of coal which appeared everywhere on the flanks of the hills and at the bottom of the valleys, and which are accessible in a degree I never witnessed elsewhere."

The visible coal that so impressed Lyell is part of a vast field that stretches along the Appalachians from Pennsylvania to Alabama. It reaches its greatest width, of some 190 miles, in western Pennsylvania near what was then known simply as the Forks, where the Ohio forks into the Monongahela and the Allegheny. Here was enough coal to build an industrial empire, but in the mid-1700s, the coal-rich land near the Forks was the site of a wetland "much infested with venomous Serpents and Muskeetose."

During the French-Indian War, the British, French, and native tribes fought brutal battles for control of the Forks, which was rich in wildlife as well as coal. When the British finally won the Forks, in 1758, a Philadelphia newspaper announced that "this valuable Acquisition lays open to all his Majesty's subjects a Vein of Treasure which, if rightly managed, may prove richer than the Mines of Mexico." The writer was not referring to the conspicuous veins of coal but to the region's wealth of beavers, needed to supply Europe's raging passion for beaver-skin hats. In fighting over what would someday be the most important industrial property on Earth, the victors were motivated by their desire to exploit a fleeting European fashion trend rather than any desire to tap into the region's coal stores.

Shortly after the British seized the Forks, they built a fort there named after their prime minister, William Pitt, and began digging the coal out of the hill across the river. Pittsburgh was born, a remote outpost on a still-turbulent frontier, but already burning the fuel of the industrial revolution. In Britain, it had taken centuries for areas to go from forested wilderness to industrial metropolis. Pittsburgh, like so much of what would become the United States, would experience that history in concentrated form, propelled in no small part by the concentrated energy beneath its hills.

The trading village of Pittsburgh grew throughout the years of the Revolutionary War, and in 1786 boasted the first newspaper west of the mountains. A local booster praised the glories of Pittsburgh, including its "vegetable air . . . constantly perfumed with aromatic flavor," and its ample fuel supplies. The author predicted, audaciously and accurately, that this tiny village some twenty days from Philadelphia by pack-horse would some day be one of the greatest manufacturing centers in the world. Within four short years, the town had progressed far enough in that direction to inspire its first recorded pollution complaint. A 1790 visitor reported that it was the muddiest place he'd ever seen "by reason of using so much coal, being a great manufacturing place and kept in so much smoke and dust, as to affect the skin of the inhabitants." At the time, Pittsburgh had 376 residents.

In the next few years, Pittsburgh truly did become a major manufacturing center. Among its earliest industries were fuel-intensive ones such as glass-making and ironworking (the settlers streaming westward needed iron tools). Pittsburgh was a lonely outpost of coal smoke on a continent that was generally relying on wood for its heat and on water for the power to animate its machinery. By 1817, Pittsburgh had a population of 6,000, and it was the largest settlement west of the mountains. It already had over 250 factories, serving markets boosted by the War of 1812. Soon the city would proudly call itself the Birmingham of America.

In the early 1800s, Pittsburgh's smoke was its most salient feature to outsiders. In 1816, a visitor reported that the town was made gloomy by the "dark dense smoke . . . rising from many parts." A few years later, another reported that the smoke and dust formed "a cloud which almost amounts to night and over-spreads Pittsburgh with the appearance of gloom and melancholy." Still another observed that when approaching the town one saw "an immense column of dusky smoke . . . spreading in vast wreaths." Walls and household items were "stained, soiled and tarnished," the inhabitants took to wearing black clothes, and, according to one report, even the skin of the locals was blackened by the smoke.

The descriptions are reminiscent of London and Manchester. We can't know how the cities truly compared as far as smoke was concerned, but Pittsburgh was clearly the smokiest city in the Western Hemisphere, and it was just getting started. By the 1860s, Pittsburgh's smoke would impress even the London-born novelist and seasoned traveler, Anthony Trollope, who would call Pittsburgh the blackest place he had ever seen, and who would sarcastically celebrate the look it gave the city:

> Even the filth and wondrous blackness of the place are picturesque when looked down upon from above. The tops of the churches are visible and some of the larger buildings may be partially traced through the thick brown settled smoke. But the city itself is buried in a dense cloud. I was never more in love with smoke and dirt than when I stood here and watched the darkness of night close in upon the floating soot which hovered over the house-tops of the city.

Pittsburgh was unique in the Western Hemisphere not just in its coal smoke and soot, but in its quick embrace of the technology that had by now transformed Britain. By the 1830s, Pittsburgh was the steam capital of the Western Hemisphere, thanks to its cheap coal.*

Although Pittsburgh shipped coal down the Ohio River, it couldn't ship it cheaply over the mountains to the eastern seaboard, where most of the nation's population and most of its factories were still located. Eastern cities got some coal from a mine operated by a British company in Virginia, opened in 1750, even before the first mine in Pittsburgh, but output was limited and costs relatively high. The cities also imported coal from Nova Scotia, and even from Britain, because it was easier to sail coal across the Atlantic than to haul it over the Alleghenies. Until a cheap source of coal was found, the eastern seaboard would use coal sparingly, and steam hardly at all. Except for water-powered textile mills, Americans would not invest in the kind of large-scale factory production by then common in Britain. Outside of Pittsburgh, the industrial revolution in the United States was on hold. But this was about to change.

IN 1790, LEGEND HAS IT, a hunter named Necho Allen was camping for the night under a protective ledge of rock amidst a

---

*In 1832, outside of Pittsburgh, almost all the largest factories of the nation were powered by water. Out of 249 factories, only four were powered with steam; more of the nation's largest industrial facilities reported using wind and mule power than steam power. In Pittsburgh, by contrast, nearly all the factories were using steam by this point. The cost of running a steam engine on the eastern seaboard was well over twice the cost of running a steam engine in England; in Pittsburgh, though, it was cheaper than in England.

region of steep mountains and dense forest in eastern Pennsylvania. The area was known then to the Indians as "the wild place," and it would be advertised later to tourists as "the Switzerland of America" because of its steep and picturesque mountains. After building a campfire, Allen fell asleep, and later awoke with alarm because, as he put it, "the mountain was on fire." He had built his campfire along an outcrop of anthracite, a form of coal so hard and shiny that it was known as "stone coal."

Like the massive field of soft, or "bituminous," coal that stretches from western Pennsylvania down to Alabama, the anthracite coal of eastern Pennsylvania was formed around 300 million years ago. As in England at the same time, Pennsylvania was overgrown with mammoth ferns and fantastic scaly lepido-dendron and other primitive trees that provided habitat not only to gigantic dragonflies and millipedes but also to our ancestral vertebrates, who were busy evolving from amphibians into reptiles. For millions of years, the shoreline periodically swept back and forth, sometimes moving inland as much as five hundred miles, as distant glaciers formed and melted. When the rising seas inundated the dense jungles, they buried them in marine sediment, leaving behind multiple layers of coal—geologic souvenirs of past climate changes.

Around 230 million years ago, the west coast of Europe and the bulge of northwestern Africa began slowly plowing into the east coast of North America, as the ocean that preceded the Atlantic closed up and the landmasses came together into the supercontinent, Pangea. As the continents collided, land on both continents was pushed upward. In northwestern Africa, the Atlas mountains arose. In North America, the Appalachians were pushed up to peaks perhaps as high as today's Himalayas.

The bituminous coal beds west of the Appalachians were somewhat buffered during this tectonic collision, but the coal beds to the east, in the anthracite region, took the brunt of it. The coal was squeezed and folded, made harder and more purely carbon, and in many places was left at steeply pitched angles. Many of the polluting impurities were forced out. Being more carbon and less hydrogen, anthracite is much harder to ignite than bituminous coal. Once started, though, anthracite burns well, as Necho Allen learned. And, perhaps more important, it burns with very little smoke, making it a far cleaner fuel than other coal—even cleaner than wood. Only a small fraction of the world's coal is anthracite, and much of the world's supply of this rare fuel is located within five counties in eastern Pennsylvania, trapped within now-eroded mountains.

The entire anthracite region was bought by the Pennsylvania government from the Iroquois Confederation for five hundred pounds sterling in 1749. Native people had been aware of the anthracite for thousands of years, apparently using it to make black paint, and even pipes. However, although the Hopis had been burning coal to make clay pots in the Southwest for over seven hundred years, there is little to suggest that the native people in the heavily forested East were using coal for fuel.

Necho Allen is just one of the people credited with discovering the region's anthracite. In 1791, a millwright discovered a particularly rich store of coal in a place called Summit Hill, near the Lehigh River. It would later be called one of the most magnificent coal properties in the world, and play a critical role in the development of the industry and the area. It included a massive outcrop of anthracite more than thirty-five feet deep that area farmers were willing to cut from the ground—quarry-style, with picks and shovels—for very low pay.

Getting this bounty of coal to market was a more daunting task, but a large and colorful cast of characters would take on the challenge. Their motives differed, but they shared a determination to do what others thought impossible, and an assumption that the future was theirs to build.

PHILADELPHIA, which in the eighteenth century was the nation's largest and richest city, is so close to Summit Hill that it can be reached by car in a couple hours or less. In the late 1700s, though, transporting coal to Philadelphia from the Summit Hill mine meant hauling it in wheelbarrows to wagons, then taking the wagons nine miles down steep mountain trails, and then traveling seven days by river, beginning with the treacherous rapids of the Lehigh.

Using the surrounding trees to build ninety-foot flat boats called arks, the mine operators sent them down the rapids with their cargo of coal and from four to six men per ark. It was by no means certain that the arks would arrive at their destination. In 1803, for example, five arks were launched down the Lehigh, but only two made it through. It was quite an adventure, though, at least according to an unattributed account of the fate of one of the arks, published a century later:

> Now the torrent roars; the waves whirl and dash madly around the boats; the men at the oars, with faces wild with animation and excitement, and with muscles full distended, run to and fro upon their narrow platforms. . . . Now the boat, shaking and cracking, swings its cumbersome form around a submerged rock; now it sheers off in a counter current towards the shore, and then bending around, again dashes forward into the rolling waves, when—cr-a-sh! je-boom!—it rises securely upon a ledge of rocks half concealed beneath the surface of the water.

The men escaped to shore, but the ark sank, leaving its cargo "to the curious speculation of the catfish and eels."

The real irony of this story, though, is that when the two surviving arks heroically delivered their product to Philadelphia, nobody wanted it; the anthracite was thrown away, except for some that was used to gravel the foot-walks. Philadelphians didn't yet know how to burn the hard-to-kindle anthracite, which requires different stoves than those that burn bituminous coal. Two days of failed attempts to make anthracite burn led one frustrated consumer to conclude that "if the world should take fire, the Lehigh coal mine would be the safest retreat, the last place to burn." (As it happened, this statement was thoroughly disproved in 1859 when a fire started in that very mine and burned, famously, for eighty-two years.) Despite this failure, in a few years there would be another bold effort to bring Summit Hill coal down to the city, this one to relieve a shortage of imported bituminous caused by the War of 1812.

To seize the opportunity presented by the war, a well-educated and connected man named Jacob Cist leased the Summit Hill mine with some partners. Instead of hiring farmers near the mine, Cist chose to bring in a labor force from his home of Wilkes-Barre, a town well north of the mine. He gathered a small but enthusiastic army mainly composed of boys from the area's most prominent families, and in July 1814 they marched off, as if to war, amidst much fanfare and singing. They had no experience in coal mining, ark building, or rapids running, but they did include a "good Latin scholar," a "natural-born poet," and one young man "principally noted for his musical powers." One of this workforce would later say of Cist and his partners, "the whole thing was a romance. Here was a chance for adventure well suited to their prolific imaginations, beside an eternity of glorious posthumous fame."

In August, they managed to send four arks loaded with coal down the Lehigh. Three of them sank. When a rock ledge tore a hole in the bottom of the fourth, it looked as if the end was near, but its crew of six quickly stripped off their clothes, stuffed them into the hole, and kept the ark afloat. These gentlemen of Wilkes-Barre arrived in Philadelphia several days later, nearly naked, bruised, blackened with coal dust, and "glad to be alive," according to Cist, who was among them. Cist and his partners were never able to make the Summit Hill mine profitable, and they went out of business after the war; but they did achieve some of the glory they sought, in part because they were able to sell this load of coal to Josiah White and Erskine Hazard, two men who would finally connect the anthracite region to the wider world.

White and Hazard made wire and nails at their Philadelphia rolling mill, and after some trial and error—and luck—they were able to figure out how to use anthracite to heat their iron. Their success with the coal shifted their focus from iron milling to anthracite. By 1818, they had won legislative approval to try to make the Lehigh River navigable. The committee chair was not optimistic that it could be done, but said, "Gentlemen, you have our permission—to ruin yourselves."

White and Hazard leased the mine at Summit Hill and raised money from Philadelphia interests. Josiah White, a prominent Quaker, took the lead. He changed out of his formal black suit into buckskins and headed for the mountains. He brought up into the Lehigh wilderness nearly a thousand laborers to build a road, create a new company town (Mauch Chunk, later renamed Jim Thorpe), and tame the river itself. White's workforce was a tough, hardbitten group mainly made up of unemployed merchant marines from various countries, and unlike Cist's crew, this one probably didn't have a Latin scholar among them.

Despite his Quaker background, White and his partners governed the workforce with military discipline. The two hundred or so men working on the river lived on specially constructed flat boats, and daily supplies of food and whisky were carefully rationed. In the maritime tradition, wasting food was punishable by what was called "cobbing," which was essentially spanking with a wooden paddle. Work stopped on Sundays to observe the Sabbath, which, according to White, his men devoted to drunkenness and violence. One partner always carried guns to protect himself from the workforce. White himself made it known that he carried no cash so that there were "no inducements upon them to commit any violence on us in the wilderness country," though it seems the spanking rule alone would have been enough to induce a violent response from those subjected to it.

Despite this less than trusting atmosphere, White and his workers, wearing special boots with holes in the toes to let the water out, managed to work together knee-deep in the cold water for three seasons a year for three years. When they were finished in 1820, they had built twelve small dams. It was finally possible to bring enough cheap anthracite into Philadelphia to assure consumers that the supply would be steady and the price competitive. The trip that had taken six days in 1814 now took one and a half. All that was left was to convince people to burn anthracite.

White and Hazard didn't just want to supply industries, which were easier to persuade, but the domestic market as well. Like other eastern cities, Philadelphia was finally facing the problem London had faced long before; it had used up the cheap, nearby firewood, and the price was rising. Still, people resisted burning coal, just as Londoners had in the late 1500s.

There were similar fears that burning coal in the home would cause health problems. Philadelphians were not worried about the smoke, and indeed burning anthracite in a stove produced much less smoke than burning bituminous in a fireplace, as the British did. Rather, they were worried about reports that the *heat* from anthracite affected the nerves, impaired the vision, caused a loss of vitality of the skin and hair, and brought on baldness and tooth decay. Gradually, these fears were alleviated, thanks in part to fairly sophisticated promotion by the anthracite coal operators. A significant number of households invested in the special stoves needed to burn the new fuel.

The work on the Lehigh River had allowed anthracite to penetrate the prized Philadelphia market, but there was a lingering problem: One could only go downstream. Arks still had to be made from upstream trees; remarkably, an experienced crew could construct an ark in only forty-five minutes, but the trees were becoming scarcer. After one trip to the city, the arks were taken apart, the hardware removed, and the wood sold. Then the ark's crew, carrying the hardware, traveled back into the mountains, on foot. The industry could not grow much more until a means of two-way transportation was found.

A FEW SHORT YEARS LATER, in 1831, the region had been transformed. As one journalist described, "From this port . . . there is a fleet of 400 vessels—a fleet more formidable than that which bore the Greeks to the Trojan War and composed of vessels, the smallest of which is almost as large as that in which Columbus ventured to cross an unknown ocean." The port in question was Pottsville, Pennsylvania, a small town in the anthracite region not far from where Necho Allen reportedly set the mountain on fire just four decades earlier.

What made this mountain village sitting a hundred miles above tidewater into a port was the Schuylkill Canal, and what most of those boats were carrying was anthracite. Opened in 1825, just ahead of the Erie Canal, the Schuylkill was the nation's first successful commercial canal—a slack-water ditch over a hundred miles long with locks, and dams, and towpaths. It was a major investment for the time, and a huge transportation advance; but, like all canals, it ultimately depended on muscle to move things. Once the canal was officially completed, boats were towed by horses or mules, although for two years before the canal opened, coal barges were pulled not by animals but by pairs of burly men pushing their chests against a board that was roped to the boat. It took them six weeks to walk the couple of hundred miles to Philadelphia and back, and the men wore out one pair of boots per trip.

The opening of the Schuylkill Canal in 1825 was just the beginning. Two years later, White and Hazard began construction on the Lehigh Canal to allow two-way navigation to their Summit Hill mine. Others cut canals across New Jersey from the Delaware River to the Atlantic, largely to carry anthracite from Pennsylvania to the northeastern seaboard. In 1825, the Delaware and Hudson Canal Company started building another canal; this one would bring anthracite out of the northern coal field to New York. The company offered $1.5 million in shares to the public, and, because canal mania was raging, it sold the entire amount in half a day from a coffee shop in New York City. The new canal was said to be the largest private undertaking in the history of the continent, shortly followed by what must have been one of the largest public bail-outs when construction costs proved higher than the company had anticipated.

Digging these canals was backbreaking work, often done by itinerant workers who moved from one canal to another. They lived in unsanitary camps where mortality rates were high. They labored with picks and shovels, and scoops drawn by horses or mules. By the 1830s, the canals they scraped across the landscape to bring coal into the cities formed the largest canal network in the nation, and America's first major inland transportation system. Canals were being built for other reasons, too, but many of the nation's other canals would be financial failures while the coal canals quickly paid for themselves as coal use multiplied. By 1859, there would be some 1,400 boats plying the Schuylkill alone, mostly carrying coal, making it by some measures the busiest waterway in the United States, even if it was just a few feet deep.

By 1829, interest in the buried wealth of the Schuylkill triggered one of the nation's first full-fledged land rushes. Because the state legislature, concerned about concentrated corporate power, would not allow the owners of the Schuylkill Canal to mine coal, the canal owners encouraged others to come into the region and open mines, thus increasing canal traffic. Into this wilderness, still known for its panther population, came thousands of ambitious speculators, most of whom knew nothing about coal mining. Half of them were teenagers. They slept lined up on barroom floors, dreaming of riches.

Most of the newly arrived left, dejected, within a few months. Some stayed long enough to set up small mines, most of which failed or sold out within five years. Still, along with the skilled Welsh and Scottish miners who were coming to the area, these pioneer operators managed to settle much of the region. New towns sprung up quickly, with names like Minersville, New Castle, and at the head of the canal, Port Carbon, which started out as one family in 1829 and grew to 912 people in a year. The anthracite region, with its distinctive, coal-driven politics and culture, was booming. Production would grow more or less steadily for over a hundred years.

The canals transported enough coal to whet the appetites of individuals and industries up and down the eastern seaboard. Soon the coal industry was no longer propelled solely by the dreams of suppliers, but by the demands of consumers, and the canals couldn't satisfy that demand for long. They were slow, they froze in the winter, and they couldn't go all the way to the mine. The answer, of course, would be railroads, which followed close on the heels of the canals, as they had in Britain.

Anthracite country is often called the cradle of American railroading, and with good reason. Apart from a couple of relatively minor exceptions, the anthracite mine operators were the first Americans to use rails, and they greatly advanced the science of building railways. White and Hazard built a nine-mile rail line from their Summit mine down to the Lehigh in 1827.

Gravity carried the coal cars and a carload of mules (who refused to walk down once they had experienced rail travel), down to the Lehigh, and then the mules pulled the cars back up.

Rails spread quickly throughout the anthracite region, and the rest of the East, and eventually locomotives followed. Schuylkill county had more track than any other small area in the country. Companies building railroads alongside the canals drained away canal business and bought up much of the region's coal property. The Philadelphia and Reading Railroad, one of the nation's largest and most powerful companies, dominated the area, but there would be five railways so intermingled with the anthracite trade that coal and railroading were often considered a single industry.

Surprisingly, these trains didn't run on coal. For the first few decades, American trains burned wood, even those that existed for the purpose of hauling coal. Anthracite did not burn well in locomotive fireboxes as they were then designed, and the bituminous fields were not yet widely exploited outside of Pittsburgh. Wood was bulky and burned fast, though, so the trains had to stop often to refuel. All along the tracks, people made money cutting wood to sell to the railroads—just one more way that trains would help turn the United States from a forested nation into an agricultural and industrial one.

The era of wood-burning trains is one that train buffs look back on with particular nostalgia. With clean-burning wood, locomotives could be much flashier: They were painted bright red and fitted with polished brass ornamentation. Engineers were flashier, too, often wearing ornate suits and vests with shiny buttons. When the switch to bituminous coal occurred later in the century, the inevitable accumulation of soot and grime meant that engines had to be painted plain black, and engineers switched to overalls. Some say that when the engineer's uniform was toned down, his status as a workman fell, too.

One problem with the shiny, wood-burning engines proved hard to ignore: They spewed out a continuous shower of sparks and cinders wherever they went, "a storm of fiery snow," as Charles Dickens called it when he visited the United States. It was a beautiful display at night, but it had a predictable downside. Wood-burning trains commonly set nearby fields and forests ablaze; some said the trains burned more wood outside the firebox than inside.

The worst problems were on the train itself, since many early passenger cars were roofless, and all were made of wood. For example, the inaugural trip of the Mohawk Valley line in New York in 1831 (just a year after the opening of the Liverpool and Manchester line) was marred when red-hot cinders rained down upon passengers who, just moments before, had felt privileged to be experiencing this exciting new mode of travel. Those who had brought umbrellas opened them, but tossed them overboard after the first mile once their covers had burned away. According to one witness, "a general melee [then] took place among the deck-passengers, each whipping his neighbor to put out the fire. They presented a very motley appearance on arriving at the first station."

Sparks on another train reportedly consumed $60,000 worth of freshly minted dollar bills that were on board, singeing many passengers in the process; according to one complaint, some of the women, who wore voluminous and flammable dresses, were left "almost denuded." Over a thousand patents were granted for devices that attempted to stop these trains from igniting their surroundings, their cargo, and their passengers; but the real cure would come later in the century, when coal replaced wood as the fuel of choice. In the meantime, some of the more safety conscious railways had their passengers travel with buckets of sand in their laps to pour on each other when they caught fire.

ALTHOUGH THE EARLIEST RAILWAYS depended heavily on wood—both for fuel and construction—the future of the railways and the industrial age would depend on iron and, later, steel. The United States could make small amounts of high-quality iron by using charcoal, but industry and railroading demanded huge quantities of cheap iron. Although coal was used to heat and work the iron after it was smelted, it generally contained too much sulfur to use for the actual smelting, when the iron ore is transformed into iron. Britain had solved this problem by making coke out of its bituminous coal, but the United States still had trouble making coke because its bituminous had too much sulfur.

As a result, Pittsburgh's industries still depended on expensive charcoal iron, and New England mainly imported coke iron from Britain. The lack of cheap domestic iron was holding back high-volume factory production in most American industries. It was also restricting the growth of the railroads, which had to depend on rails shipped across the Atlantic.

The key to modernizing the American iron industry would be anthracite, which is naturally low in sulfur, but it took special technical innovations to make it work. It wasn't the iron industry that pursued the innovation, but coal producers looking for a huge new market. In the late 1830s, White and Hazard persuaded a Welsh ironmaster named David Thomas, who had succeeded in smelting iron from Welsh anthracite, to come to the United States and build an iron furnace near their mine. Others copied them, and by 1849 there were sixty anthracite furnaces making iron in eastern Pennsylvania and twice as many five years later. The nation's iron output multiplied and the price plummeted.

Within a few years, two major bottlenecks constraining American industrialization—the lack of cheap coal and the lack of cheap iron—had been eliminated by the products of this small corner of Pennsylvania. The stage was set for a burst of economic development.

With ample coal and iron, steam power spread quickly, and it began to change the nation in the same ways that coal had begun changing Britain decades earlier. The opening of the anthracite fields and the modernizing of the iron industry led to the rise of mass production between 1835 and 1855 in many American manufacturing and mining industries. Small enterprises, powered by wood machines and water power, were on the way out. Large factories employing iron machines and coal-fired steam power were on the rise. Even industries that had produced on a large scale in the age of wood, such as the textile industry, expanded as they turned to coal fuel and iron machinery.

The locus of industry shifted from small towns to urban centers, where wage laborers could be found easily. By the end of this period, the United States was producing manufactured goods that even the British admired. The country was still largely a rural and agricultural nation, but it was changing quickly. In the 1840s alone, manufacturing rose from 17 percent of national output to 30 percent. As mines, ironworks, factories, and railroads grew larger, new classes of managers and laborers emerged, and major corporations became an increasing presence in American life.

The American Northeast's coal-fired economic surge before 1860 further deepened the political and economic divisions between the industrial North, dependent on coal-burning factories, and the agrarian South, dependent on slave-exploiting plantations. It also ensured that when the Civil War finally came, it would explode with an industrial intensity. Failing spectacularly to live up to early British expectations that the railways would bring about a brotherhood of man, trains brought hundreds of thousands of troops and a constant stream of munitions to the battlefields, enabling larger and bloodier battles. And industrialization ultimately ensured a Union victory. The North had a decisive industrial advantage over the South with ten times more factory production, fifteen times more iron, thirty-two times more firearms production, and, most dramatic, a 38-to-1 advantage in coal.

After the Civil War, the industrial interests of the North were able to devote their attention to the West—to extending the railroad to the Pacific, literally linking the nation together with iron bands. Those trains were not yet burning coal, but they ran on rails formed with coal, were pulled by engines made with coal, and were financed by empires built on coal. The trickle of westward pioneers who had braved the wagon trails turned into a flood of settlers riding the rails. Now the door was open to ranching and farming, because the surplus could be shipped by rail to the urban markets of the East. America's push across the continent had begun long before the industrialization of the East, as had the decimation of native tribes. But the added power of the East's firm industrial base dramatically accelerated the taming of the wilderness and the defeat of its native inhabitants.

Coal-fired industrialization changed the nation in another way, too. The rise of large industry in the United States was a stark challenge to the nation's image of itself as an egalitarian land of farmers. America, or at least those Americans most attached to the Jeffersonian ideal, celebrated the citizen-farmer as the solid foundation of representative government, his power firmly grounded in the small parcel of land he worked. To many, the factory was simply un-American, a form of evil inextricably linked to dangerous concentrations of power, British class hierarchies, and repression. In the words of one textile worker, "Manufacturing breeds lords and Aristocrats, poor men and slaves. But the Farmer, the American Farmer, he, and he alone can be independent."

The country was big enough that this agrarian ideal could live on in many places within its borders, alongside the ideal of pioneers transforming the wilderness. In the nation's coal regions, though, and in the industrial cities that now depended on them, the "precious seed" was yielding a new kind of America, with different goals, different challenges, and very different politics than the one it was replacing.

A primal forest from the Carboniferous, or coal-forming period, about 360 to 290 million years ago, dominated by trees like the towering *lepidodendron* (right) and the *sigillaria* (center and left). *Courtesy of Peabody Museum, Yale.*

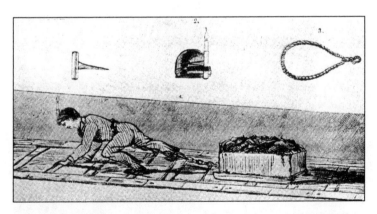

Illustration from an 1840 British government report on child labor, showing how children were forced to crawl through narrow mine tunnels hauling coal.

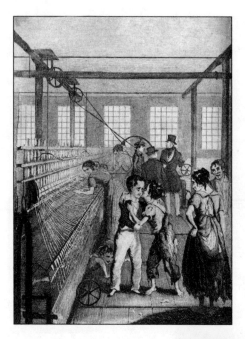

Illustration of a cotton mill from an 1840 British novel, seeking to draw attention to the miseries of the nation's growing class of child workers in its coal-powered factories.

Some of the dark and confined industrial housing hastily built in the early nineteenth century for workers in the new coal-fired factories of Britain's industrial cities.

The effects of the bone-softening disease rickets, to which children growing up in the smoke-filled slums were particularly prone due to lack of sunlight.

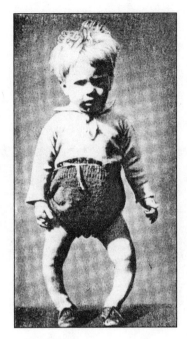

The historic Liverpool and Manchester Railway, which took railway technology out of the coal fields and launched it into the public consciousness. *Courtesy of Yale Center for British Art.*

The Lehigh Canal, which along with other canals made coal a significant industry in America and prompted a surge of industrialization. *Courtesy of National Canal Museum.*

The Corliss Steam Engine, which through 8 miles of connecting shafts powered the thousands of machines showcased at the nation's Centennial Exhibition in Philadelphia, 1876.

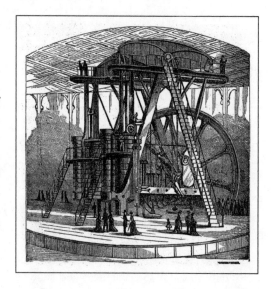

"Breaker boys" at work in 1911, picking stone and slate from the coal as it rushes beneath them on a conveyor belt.

Women at a coal stove. One study found that coal stoves demanded on average one hour a day of stoking, cleaning, and polishing. *Courtesy of Wisconsin Historical Society.*

An image taken at 3:00 in the afternoon on a particularly smoky day in Pittsburgh, 1913. *Courtesy Carnegie Library of Pittsburgh.*

1909 cartoon mocking the offer by a Chicago women's club to assist the city smoke inspector by watching for non-compliance with early smoke abatement laws.

A coal-fired power plant formerly located on the banks of the Mississippi River in St. Paul. Its towering smokestack was demolished with explosives before a cheering crowd in 2008.

A small coal mine in China. In addition to its large and relatively modern coal mines, China still has tens of thousands of small ones.

One of thousands of carved Buddhas at a site in Shanxi province, which sit above the richest coal field in China. The fifteen-century old carvings are threatened by China's dense air pollution.

six

The Rise and Fall of King Coal

On May 10, 1876, the United States launched its first centennial celebration at the grand Centennial Exhibition in Philadelphia. In the months ahead, 9 million visitors would make the pilgrimage to marvel at the material wonders spread out along the vast fairgrounds. One of the most popular attractions was the behemoth of a coal-fired steam engine that powered the thousands of mechanical inventions set up in the exhibition's sprawling Machinery Hall. At the opening ceremony, not long after the choir sang Handel's "Hallelujah Chorus" to an exhilarated crowd of 100,000, President Ulysses S. Grant stepped up to the giant engine and pulled its control lever. With a hiss of steam, the engine's cross beam began to rock, its two enormous pistons began to churn, and eight miles of connecting shafts running through the hall began to rotate. A cheer rose from the crowd as fourteen acres of gleaming machinery simultaneously sprang to noisy life, spinning, sawing, sewing, pumping, and printing, trumpeting in unison the nation's arrival as an industrial power.

Historians have seen in the Centennial Exhibition a vivid symbol of the emerging industrial era in the United States, a showcase of the nation's optimism, its inventiveness, and its love of technology on a massive scale. Played out across the country, these same factors, along with a flood of cheap labor and plenty of coal, would soon make the United States the world's leading industrial nation. As in Britain, this new power would have dramatic effects on working and living conditions, and on the air. In the United States, though, still in its formative stages, coal would have an even greater impact on the political power structure of the nation. Nowhere was that more obvious than in the anthracite communities just up-river from Philadelphia's exhibition, where the criminal trial of the Molly Maguires was simultaneously showcasing a very different side of the nation's growing strength.

THE MOLLY MAGUIRES were allegedly a secret organization of Irish Catholic coal-mining terrorists who for many years advanced their interests in the anthracite region through arson, beatings, and the systematic murder of coal bosses and others who stood in their way. The Mollies had their roots back in Ireland, where secret societies had long battled British oppression. Molly herself, according to legend, strapped two pistols to her thighs and led groups of men disguised in dresses through the Irish countryside to conduct nighttime attacks on the British. When the Irish, fleeing famine, came in droves to the anthracite region of Pennsylvania, they found themselves subject to a familiar oppression, and the legend of the Molly Maguires rose again. By May 1876, they were blamed throughout the nation for plunging the anthracite region into violence. Their notoriety was thanks largely to the work of one man, Franklin B. Gowen—railroad president, union-buster, cartel-builder, and crime-fighter.

Gowen began his career apprenticing with a coal trader, and at age twenty-three spent a year or so losing his savings by trying to operate a coal mine in anthracite country. After that failure, he switched his focus from coal to law, serving briefly as Schuylkill County's district attorney, and in 1867 moved to Philadelphia to become legal counsel to the already mighty Reading Railroad. With a personal forcefulness and magnetism that made him, in the words of one historian, "well-nigh irresistible even when his arguments were obviously preposterous," Gowen was running the railroad within two years, at the age of thirty-three.

Gowen was an early champion of the "gospel of bigness" for American industry. Central to his vision, and to the Reading's profits, was a cheap and steadily growing coal supply, and this Gowen tried to ensure by controlling both the unionization and the fierce competition that he blamed for making the industry unstable. First, Gowen indirectly fought the miners' newly formed union by raising freight rates to coal operators who threatened to give in to the union's demands. He could control the unions much more effectively, though, if the Reading owned the mines directly, so he slipped a law through the state legisla-

ture letting the Reading mine coal even though its original charter had explicitly forbidden it to do so. He soon spent 40 million
dollars buying up most of the coal mines in the Reading's portion of the anthracite region.

Then, rather than compete with the coal production from
the other two portions of the anthracite district, production
that was controlled by the other railroads and a handful of independent mine owners, Gowen brought them together into a
price-fixing agreement in 1873. This "pool," as it was called, was
the nation's first major industry-wide cartel. (It is no accident
that the classic board game Monopoly includes the Reading
and other coal railroads). Neither secret nor at the time illegal,
the pool was reported in the New York City papers the next
day. After just four years in office, not only had Gowen managed
to make anthracite the most organized industry in the nation,
but he had also gravely weakened the miners' efforts to organize
themselves.

The image of a tyrannical King Coal whose power extended
far beyond the coal camps was starting to form in the public mind.
By encompassing nearly every lump of anthracite in the nation,
the cartel reached into the hearth-fires of millions of Americans. In
1875, when Gowen and the coal operators cut wages and the miners went on strike in response, the public sympathized with the
miners. Newspapers that normally condemned all strikes now
denounced the coal cartel that "with one hand reaches for the
pockets of the consumers, and with the other for the throats of the
laborers." The strike, lasting five long months, was marked by vio-

lence on both sides. Striking miners were beaten and killed, as were strikebreakers and mine bosses. Miners derailed trains, sabotaged machinery, and burned down mine buildings. The newly combined coal operators held firm, though, and ultimately the hungry miners straggled back to work at the lower wages, their union essentially destroyed. The miners blamed Gowen, and for years they did not speak his name without a curse.

It wasn't long before the Pennsylvania legislature began to investigate Gowen's monopolistic strategies. Appearing in person before the investigating committee, Gowen persuasively argued that large mining companies were in the public interest because only they could make the needed investments. Then he quite effectively changed the subject: He read out a long list of threats, beatings, fires, and shootings committed by "a class of agitators" among the anthracite miners. When he was through, the focus of the legislature and the public (for Gowen published his arguments) had shifted from the Reading's growing power to the region's growing wave of organized crime.

Gowen's list of crimes had been compiled by Allan Pinkerton's private detective agency, which Gowen had secretly hired two years earlier to infiltrate the Molly Maguires. Pinkerton had sent an Irish Catholic spy into the region, and after he had gathered evidence of their crimes, and perhaps provoked additional ones, the trap was sprung. In September 1875, scores of suspected Mollies were rounded up by the Coal and Iron Police, a private security force which was controlled by Gowen and was the main law enforcement agency in the region.

The following spring, a spectacular and high-profile murder trial of five of the suspects opened in anthracite country. Not only did Gowen's secret agent testify against the suspects, who had been arrested by Gowen's private police, but the prosecution team was led by none other than Gowen himself, the former district attorney now acting as special prosecutor for the state. It would be hard to find another proceeding in American history where a single corporation, indeed a single man, had so blatantly taken over the powers of the sovereign.

Gowen, ever flamboyant, appeared in the courtroom dressed in formal evening clothes. Before an electrified audience, he presented a case not just against the five suspects but against all the Molly Maguires, and, by strong implication, against the miners' now-defunct union. At issue was not just the murder with which the suspects were charged but a whole array of crimes. Following Gowen's line of reasoning, the press soon blamed the Molly Maguires for all the labor violence by miners during the long strike of 1875. After a series of trials, twenty accused Mollies were hanged, and twenty-six more imprisoned. For bringing down the Mollies, Gowen—so recently the subject of public scorn and suspicion—was lauded in the press for "accomplishing one of the greatest works for public good that has been achieved in this country in this generation."

Two conflicting lines of folklore have emerged around the Molly Maguires, one branding them brutal criminals, the other hailing them as martyrs in the battle against King Coal and corporate tyranny. Modern historians generally agree that the legend of the Mollies was greatly magnified by Gowen's oratory and by the press, and that the wave of crime against coal producers in the area, particularly after the long strike of 1875, was the predictable result of the miners' desperation rather than the work of a structured secret society. Clearly, the miners' union, far from being dominated by the Mollies, had helped prevent violence by the miners while it existed. In the public's mind, though, organized anthracite miners were now seen as terrorists, and support for miners' attempts to unionize withered away. The specter of the Molly Maguires so completely undermined subsequent attempts to unionize that no union would succeed in organizing the anthracite miners until the United Mine Workers did so at the end of the century.

Gowen's battles were not over. He had put the Reading disastrously into debt buying so many coal mines. Most of the Reading's stocks and bonds were held by powerful London financiers, and they were not happy. Even so, Gowen kept trying to expand, and while he believed in cooperation among coal producers, he was fiercely competitive with other railroads. When he tried to establish competing rail lines to New York and Pittsburgh, in alliance with William H. Vanderbilt, Gowen provoked one of the last great wars between the eastern railroads and made powerful new enemies.

Ultimately, the overextended Reading was unable to pay its bills, and a New York banker of rising influence named J. Pierpont Morgan was brought in to reorganize it. Morgan already held interests in many of the eastern railroads and had strong links to London financial circles. His approach to railroads was like Gowen's approach to coal: He wanted to control what he saw as wasteful competition. Morgan didn't approve of Gowen's bold moves to expand the Reading's reach, and in 1886 he forced Gowen, that pioneer of anti-competitive business methods, out of the Reading for being too competitive.

The evening after he managed to push Gowen out, Morgan went to a family party. His son reported that "Papa is simply triumphant about this Reading business," and feeling especially "bright and well." Gowen went back to the practice of law; but three years later, on a Friday in December, he locked the door of his hotel room and shot himself. For twenty years he had been one of the most talked about business leaders in the United States, as well known in his day as Andrew Carnegie or John D. Rockefeller. Although not primarily a coal mine operator, he had done more to establish the national image of the far-reaching King Coal than anyone of his era. In a few decades, Gowen would be largely forgotten by history, his name sometimes appearing misspelled in footnotes listing him merely as a prosecutor in the trial of the Molly Maguires.

Although a few independent anthracite producers built private family fortunes in the region, the vast majority of the nation's anthracite production continued to be controlled by the Reading and a handful of other railroads for decades more. Because anthracite was the subject of the cartel and endless labor battles, the image of King Coal and the coal barons fixed itself in the public's mind, even though anthracite production was actually in the grip of the railroads, which were in turn controlled by financiers led by Morgan. Anthracite formed the bottom of an economic food chain that channeled energy throughout the regional economy, and that drew profits away from the coal communities to the power centers of Philadelphia, New York, and, to a surprising extent, London. After having built an important link in that chain, Gowen was simply replaced at the top of it by an even bigger fish, one who would remain there for a very long time.

IN 1876, THE NATION still got twice as much energy from firewood as from coal, but wood's share was dropping fast. Coal consumption doubled every decade between 1850 and 1890. In the late 1890s, the United States became the world leader in coal production, finally bypassing Britain. (Germany, that other rising industrial power, held third place.) By 1900, coal was the unrivaled foundation of U.S. power.*

Although the anthracite fields still grabbed more headlines, after the Civil War the bituminous fields provided more coal as demand from the growing industries and cities of the Midwest skyrocketed. By 1900, the nation would burn nearly four times more soft coal than hard, in part because the soft coal was much more widely distributed. In addition to the rich seam running through the Appalachians from Pittsburgh to Alabama, there was an important bituminous field centered in Illinois and Indiana.** Unlike anthracite, the far-flung soft-coal

---

*Coal provided 71 percent of the nation's energy; oil, natural gas, and hydro power each provided less than 3 percent, and wood was down to 21 percent.

**While most bituminous came from these areas, the nation's wealth of soft coal was so spread out that it would ultimately be dug from over 9,000 mines in thirty-three states.

industry did not lend itself to monopoly or oligopoly control. Soft coal, in fact, became one the nation's most fiercely competitive industries.

This competition among the owners gave the bituminous miners one big advantage over the anthracite miners. The anthracite owners worked as a team to destroy the miners' short-lived unions, but the bituminous owners could not. Indeed, some owners in the midwestern coal fields actually saw a miners' union as the only force that could stabilize the overly competitive industry, particularly if it could unionize the competing coal fields. In this midwestern field in 1898 a union of miners won a historic regional contract granting them recognition. It was the first coal miners' union to gain a lasting foothold in the United States, and it would soon become the largest and most powerful union in the nation: the United Mine Workers (UMW).

At the head of the UMW stood a young idealist named John Mitchell. At age twelve, Mitchell had begun laboring in the mines, acquiring a permanent stoop. In a few years he left the mines and became a union organizer, walking from coal camp to coal camp in southern Illinois. In 1898, at age twenty-eight, Mitchell became head of the newly recognized union and looked to expand its reach to competing coal fields. He took on anthracite country, even though no union had been able to take root there since the Molly Maguire trials a generation earlier.

The longer hours and tighter controls on the workers made conditions in the anthracite fields worse than in the midwestern field. Miners were more likely to live in company houses from which they could be evicted for union sympathies, and were more scrutinized by company spies. Child labor was even more rampant. Above-ground, thousands of so-called breaker boys did the backbreaking job of sorting the coal by sitting suspended over conveyor belts and reaching down between their legs to remove the slate and stone as it rushed along. Other boys worked below ground, often guiding the mules that hauled the coal. (The mules were generally stabled underground, often going years without seeing the sun or the grass.)

The anthracite mines were deadlier, too, with hundreds killed yearly by cave-ins, explosions, gases, and floods. Although larger accidents killing a hundred or more prompted safety legislation, ongoing threats, such as small cave-ins, attracted far less regulatory attention, and miners turned to other safety measures, such as befriending the mine rats. Notoriously bigger, meaner, and uglier than surface rats, mine rats were thought to sense subtle shifts in the mine workings; if they suddenly scurried away, the miners followed. This was true in bituminous mines, too, where one observer wrote, "It is a common sight to see a miner feeding half a dozen or more [rats] from his dinner pail. Frequently they become so tame that they will climb on a miner's lap as he sits at his lunch and crowd around him to receive such portions of his meal as he has taught them to expect."

This was the life to which thousands of hopeful immigrants had flocked, including a new wave of Slavs, Hungarians, and Italians, some lured there by the railroads through European labor agents. The new arrivals were disliked by the English-speaking workers because they worked harder for less; but Mitchell and the UMW made a concerted effort to reach them, and even hired organizers who spoke their languages. The southern and eastern Europeans were soon so intensely loyal to Mitchell that when word spread that "the president" had been shot in 1901, they gathered in crowds, weeping. When they were told that it was only McKinley, not Mitchell, they went home greatly relieved. Despite what many had believed when these immigrants arrived, with the right leader they were willing to fight against the dehumanizing conditions they were forced to live and work in.

In May 1902, the fight began. Mitchell launched a show-down between what was already the nation's biggest union and its most powerful industrial combination over its most vital commodity. Nearly 150,000 anthracite miners went on what was later called "the best managed of any strike that ever occurred in the United States." The bituminous mines kept producing, but as the nation shifted from hard to soft coal, it caused a bituminous shortage as well. The United States faced a true energy crisis, but instead of automatically denouncing the strikers, the public, the press, and the politicians generally sympathized with them, won over by the justice of the miners' cause and by Mitchell's nonmilitant style. In the words of Mitchell's biographer, Craig Phelan, the public saw Mitchell as "the prince of moderation, a progressive who called for a measure of justice and not revolution, a man who stood ready to arbitrate any dispute."

As the strike dragged on, Theodore Roosevelt invited Mitchell to the White House to negotiate a settlement with Gowen's successor as president of the Reading, George F. Baer. Baer's lack of public relations skills had already won him the nickname George "Divine Right" Baer, bestowed upon him after he argued that the workers' welfare would be protected by the men whom God had entrusted with control of the nation's property interests, not by labor agitators. At the White House, Baer managed to so offend the president that Roosevelt said he nearly "chucked him out of the window" for his arrogance.

Ultimately, Roosevelt went over Baer's head and pressured J. P. Morgan, still in control of the railroads, to settle the strike. The miners received only a 10 percent wage increase, but the settlement was hailed by most as a union victory, and a well-deserved blow to the coal barons. A landmark in U.S. labor relations, it was the first time a president had intervened in a major labor dispute on the side of the workers, and it made the leader of the coal miners a national figure whose peaceful and conciliatory image provided, in Phelan's words, "a new model for union leadership."

The 1902 strike was also a vivid lesson in how dependent the nation was on coal, and how deeply it could be hurt when supplies ran short. Even after the strike was settled, it was months before coal supplies and prices were back to normal, and some communities experienced great hardship. In January 1903, three hundred citizens of the town of Arcola, Illinois, politely mugged a coal train that broke down there on its way to Chicago. No coal had been delivered to Arcola for a month, half its citizens were out of coal, and businesses had been closed for a week. When the railroad refused to sell its coal to the town, the good citizens simply surrounded the train with their wagons, climbed up onto the coal cars, and began shoveling. Active members of the raid included the town's two bank presidents, two ministers, and a police officer. One of the bank presidents kept a careful accounting of the amount taken by each person so that payment could later be made.

CLEARLY, THE NATION'S coal supply was no abstract political issue but one close to people's hearts. In cities and towns throughout much of the nation, millions now depended on it to heat their homes and cook their meals. Many rural households could still heat with wood, but even there the increasing use of stoves was making wood less popular since it was harder to chop into the small sizes required by the controversial appliances.

As in Britain, stoves were initially embraced with reluctance. They were disliked for the same reason: They hid the fire. To many, an open flame was central to domestic comfort. It was the warm memory of the open hearth's "roaring, hilarious voice of invitation, its dancing tongue of flames," that inspired the Revolutionary soldiers as they struggled through wartime hardships, claimed Harriet Beecher Stowe, suggesting they would never have gone "barefoot and bleeding over snows to defend air-tight stoves and cooking-ranges."

Ultimately, though, fuel efficiency won out over tradition and beauty. Fully enclosed cast-iron stoves began to spread quickly in the 1830s, and were in common use after the Civil War.

In many American homes, the open fire, the flickering light that had been the daily focal point of our species' domestic life since before we were fully human, disappeared into a box. (Perhaps its loss helps explain why a substitute flickering light, in the form of television, would be so warmly embraced a century later.) The move away from open fires brought another important change into the home as well: cooking, which since time immemorial had forced people to stoop over a fire, could finally be done standing up.

The 1897 Sears Catalogue offered a variety of stoves priced from $5.97 for a simple one to $48.00 for the fanciest model. They were made of cast iron, though they would later be replaced by stoves made largely of rolled steel and lined with sheets of that handy household insulator, asbestos. Of course, some stoves were designed simply to heat rooms, and lacked a cooking surface, but kitchen stoves had openings directly above the firebox so that properly sized pots could sit over the flame. Located below or beside the firebox was an enclosed oven that was heated by closing a damper and sending the hot exhaust from the firebox down and around the oven before it could escape up the stove pipe. The stove could be fitted to burn wood, coal (hard or soft), or both. Coal was in many ways an easier domestic fuel than wood; it didn't have to be chopped, there was less weight to haul for the same amount of energy, and it created a longer-lasting, steadier fire. By modern standards, though, the coal stove was anything but easy.

Just lighting the stove was a complex affair, and home economics texts and women's magazines included plenty of advice on how to do it properly. First, the ashes from the last fire had to be removed from the stove's firebox—but not all of them, since a little half-burned coal at the bottom was useful. Because coal is not easily lit, greasy paper might be put into the fire box, along with wood shavings and then larger strips of wood, criss-crossed to allow airflow but to keep coal from falling through. A little coal could then be added, but not too much. An updraft in the chimney pipe might need to be started by thrusting a burning wisp of paper into it. Minutes later, after the kindling and first scoop of coal had taken light, more coal could be put in. Ashes had to be periodically shaken loose, and more coal added (the lumps carefully arranged by a gloved hand, according to one magazine). The dampers and vents also had to be continually adjusted to control air flow and the speed of combustion. Stoves were often kept burning all day, and sometimes all night, to avoid the bother of rekindling, to keep a reservoir of warm water ready, and to keep the kitchen warm.

In 1899, an experiment at Boston's School of Housekeeping found that nearly an hour a day was needed just to tend to a modern coal stove. Most of that time went to the basics of maintaining the fire—carrying the coal (292 pounds used over six days), setting and stoking the fire, and sifting and emptying the ashes—but a third of the time was spent rubbing black-lead onto the stove's surface to keep it from rusting. "Black-lead" was the popular term for graphite, and graphite, or at least much of it, is made when the "coalification" process goes on too long, squeezing and heating coal so intensely that everything but the carbon is driven away.* Graphite, like a diamond, is pure carbon but in a softer form, useful in many humble ways, including the making of pencils and the prevention of rust.

So it was that people spent hours every week tending to the coal burning inside their stoves, and then spent hours more carefully rubbing an over-processed version of coal on its surface. These hot, black, soot-covered, carbon-smeared monsters sat at the heart of millions of homes like giant sculpted pieces of coal, endlessly demanding and hard to control, but able to radiate the energy of the Carboniferous into the kitchen. This radiated heat was considered a blessing for about eight months of the year—the kitchen coal stove was often the home's only heat source—and a curse during the rest. One problem, of course, was that a solid-fuel stove could not simply be turned on when needed and off when the cooking was done; it took a long time to get the stove hot enough to cook, and then there was no easy way to put out the fire without wasting the fuel within it. Early twentieth-century home economics books recommended that those who could afford it buy a separate kerosene stove to use in the summer months.

In larger homes, more coal was burned in parlor and bedroom stoves. In 1869, Catherine Beecher and her sister Harriet Beecher Stowe estimated that in the mid-Atlantic states, it took

---

*I opened a fortune cookie recently that incorrectly informed me that "a diamond is a piece of coal that stuck with the job." This cultural myth may be from the comics; when Superman squeezed coal in his fist, he made diamonds. When nature squeezes coal, it makes graphite.

three tons of anthracite to keep one fire burning through the winter, and four tons in the northern states. The new practice of central heating meant that some of the large new homes were also burning coal in the basement. The emergence of forced-air furnaces was not well received by the Beecher sisters, who believed that furnaces were a means by which uninstructed housekeepers would "not only poison their families with carbonic acid [carbon dioxide] and starve them for want of oxygen, but also diminish health and comfort for want of a due supply of moisture in the air."

Indeed, there was increasing concern about indoor air quality, and it was connected not just with furnaces but with stoves generally. It was widely recognized that leaking stoves could poison a home with the truly deadly carbon monoxide, but the Beecher sisters were describing something else—a more subtle kind of poisoning by "vitiated" air. It was feared that stoves made the air unhealthy because, like people, fires consume oxygen and release carbon dioxide. Fears of oxygen depletion merged in a confused way with the not-yet-abandoned miasmic theory of disease and the developing germ theory. In combination, these theories made stuffy rooms seem like death-traps and assigned to the homemaker the duty to keep her family healthy by keeping the windows cracked open. Air quality had become not only a health issue but also largely a women's issue, and it would remain so for decades.

There was one other common use for coal in the home: lighting. When the Beecher sisters wrote in the late 1860s, gas lighting had already spread from England into the cities and large towns of the United States. This was not natural gas, but gas made from coal. The coal would be baked at the gasworks on the edge of town to drive off the gases, which were then piped beneath the streets and into the street lights and people's homes. The gas pipes ran through the walls and ceilings and into light fixtures controlled by a small key that could turn the gas flow on and off (with fatal consequences if these keys didn't work properly). Eventually, the fixtures had incandescent mantles placed over the gas jet to make them glow more brightly than a naked flame. The mantles were then generally covered by glass globes, and the fanciest chandelier-like fixtures were sometimes known as "gasoliers."

Where gas was not available, the main alternative source of light after the Civil War was kerosene. Kerosene was the first petroleum product to find a wide market, allowing Standard Oil to build an industrial empire long before the rise of gasoline and the internal combustion engine. These fossil light sources helped put an end to the use of sperm whale oil. Coal and oil thus helped save the whales, just as coal had for centuries helped save the remaining forests.

The widespread use of gas lighting was important for other reasons as well. People got used to the idea of obtaining their fuel "from the outside"—on demand from community-wide distribution systems—instead of from their own privately stored fuel stocks, and local governments began to see energy distribution as a valid municipal function. People also grew comfortable with gaseous fuel, something new to human experience. In these ways, the gaslight era helped pave the way for the electricity and natural gas networks that would provide most household energy in later years. Gas lighting was also the first coal product to compete directly with oil for the nation's energy dollars. The use of gas lights spread with the growth of cities, making kerosene increasingly just a rural product. Coal won the lighting market more definitively when both gas lighting and kerosene were replaced by electricity, almost all of which was made with coal. Electricity would be coal's only major victory over oil.

THE 1902 STRIKE served to emphasize how the nation had divided into clean anthracite cities like New York, Philadelphia, and Boston, and dirty bituminous ones, like Pittsburgh, Chicago, St. Louis, Cincinnati, and Birmingham.* In New York City, as the strike-induced shortage caused anthracite prices to rise, more and more coal users turned to bituminous, violating city laws and alarming city residents. Some plants allegedly switched to bituminous after dark when the smoke would attract less attention. In June of that year, the *New York Times* carried the distressing headline "Smoke Pall Hangs Over the Metropolis," something that was true every day in the soft-coal cities. A

---

*In fact, nearly every city west of the Appalachians depended heavily on bituminous except the emerging cities in Texas and California, where oil and natural gas were locally available. The reason was the cost of transportation; bituminous was mined in some twenty states by 1900, anthracite only in eastern Pennsylvania. In St. Louis, for example, anthracite could cost four times more than soft coal from the mines of Illinois.

letter to the editor, bemoaning the growing illegal use of bituminous during the strike, asked, "Are we to have fastened on us the frightful infliction which curses Pittsburg and Chicago?" Andrew Carnegie, the steel magnate whose bituminous-burning mills in Pittsburgh added greatly to that city's "frightful infliction," chose to live in New York City, and warned: "If New York allows bituminous coal to get a foothold, the city will lose one of her most important claims to pre-eminence among the world's great cities, her pure atmosphere."

Pure atmospheres were something most bituminous cities had not seen, and would not see, for a very long time. Most of the bituminous cities saw their coal smoke as the inevitable byproduct of industry, and saw industry as the source not merely of their material wealth but of modern civilization itself. And yet, this belief was not unshakeable. By the late nineteenth century, it was starting to clash with other firmly held attitudes that linked civilized life to cleanliness, beauty, health, and ultimately morality. In short, coal smoke was coming into conflict with an emerging environmental philosophy.

This was also about the time that Americans were cultivating a heightened appreciation for nature in its wildest state. But the growing antipollution movement emerged not from the forests but from the kitchens. The nation's smokestacks were under attack by the nation's women, who took the domestic cleanliness principles they had for so long used to protect their families' health beyond their front steps and into the community at large in what was called the "municipal housekeeping" movement.

All around the nation, middle-class and upper-middle-class women with the leisure to devote to activism joined civic clubs like the Ladies Health Protective Association of Pittsburgh and the Wednesday Club of St. Louis. Smoke was not the only kind of urban filth attracting the attention of the municipal house-keepers. Indeed, "smoke abatement" was part of a much larger reform movement focusing also on water supply, sewage disposal, and solid waste removal. Solutions to these other problems, while not cheap, were relatively simple: huge, publicly funded infrastructures. Citizens and businesses just had to pay their tax bills, and then take advantage of the new amenities. With coal smoke, governments did not offer a publicly financed solution, but tried to do something much harder: persuade those putting the smoke into the air to find and finance their own solutions.*

The women's clubs made many alliances—with the medical profession, with engineering clubs, with business leaders, and with other civic clubs—but they very much led the charge. In his recent analysis of this era, environmental historian David Stradling concluded that "given women's roles as keepers of the house and protectors of morality, female voices decrying the ravages of impure air echoed deep into society." But because they were women, and often women of privilege, their desire for

---

*However, in the mid-1800s, a "gaseous sewage" system did gain a few proponents in England. They suggested that the coal smoke of every home be vented into the sewers along with the liquid waste and carried away, to be released through immense chimneys a safe distance from town. This idea, possibly fanciful and seemingly dangerous, never caught on.

cleaner air was often seen as frivolous and sentimental. Women in Pittsburgh were accused of wanting to abate smoke to protect their complexions rather than to protect health.

In their search for solutions to the threat of coal smoke, most of the women's groups actually took a highly unsentimental and practical approach. They worked with engineers to find technologies and fire-tending practices that would let fires burn more cleanly. They were still unable to vote, but they successfully lobbied for new municipal laws banning dense smoke* and worked to ensure the laws were enforced. Unfortunately, their enforcement efforts were not always taken seriously. When club members in Chicago offered to help the city smoke inspector identify violating smokestacks in 1909, the newspaper printed a cartoon that showed women in flowery hats perching delicately on city rooftops and watching smoke pour from chimneys while they stitched their needlepoint.

It was generally understood that merely aesthetic objections to smoke and soot were insufficient to warrant interference with something so vital to the nation as coal burning, so the impact of smoke on health became the focus of most activists. Unfortunately, the science of measuring the smoke's health effects had not progressed much beyond John Graunt's feeble attempts centuries earlier. The belief that smoke had antiseptic properties still lingered; indeed, the era's tight focus on germs and epi-

---

*There were many legal set-backs as major coal-consuming businesses challenged the legality of the ordinances, but by 1916 some seventy-five cities had adopted smoke abatement ordinances.

demics, which had motivated cities to spend vast sums on water
and sewage projects, made smoke seem safe by comparison. As
late as 1913, when Birmingham steel mills pressed the city to
repeal its new smoke abatement law, a physician supported their
case, pointing out that smoke could not possibly be harmful
because, having been purified by fire, it did not carry germs. A
Chicago coal dealer defending against smoke abatement efforts
had gone even further when, in 1892, he argued that the black
carbon deposited by smoke in the lungs actually purified the air
as it passed through the carbon and into the blood.

Of course, those pushing for smoke abatement saw the issue
quite differently. Some of the early activists blamed coal smoke
for causing not just deadly lung diseases but also a variety of
other problems, including acne, diarrhea, and constipation. In
1905, the American Medical Association blamed smoke for making children "pale and flabby." Some went further, accusing the
"smoke evil" of promoting moral depravity. In 1909, the president of one of Chicago's women's clubs claimed that "Chicago's
black pall of smoke, which obscures the sun and makes the city
dark and cheerless, is responsible for most of the low, sordid
murders and other crimes within its limits. A dirty city is an
immoral city, because dirt breeds immorality."

Some members of the medical profession, while not going
that far, shared the belief that smoke seriously affected mood and
behavior. A physician writing in 1913 asserted that women living
in gloomy, sunless homes were likely to "be irritable, to scold and
whip their children and to greet their husbands with caustic
speech." The husbands, in turn, were driven to drink, and the

children became "dull, apathetic, and even criminally inclined." A psychologist investigating the effect of smoke in Pittsburgh that same year found among its residents a "chronic ennui" and depression; he also believed that the dense atmosphere made "clear, trenchant, reflective thinking" more difficult for him during his time in the city. The extreme assertions blaming most urban crimes on smoke are easy to dismiss, but the underlying question about smoke's impact on mood and behavior remains an interesting one, particularly given more recent findings linking light-deprivation to depression.

In time, the scientific debate on smoke's health effects focused on lung disease. In the late 1800s, pulmonary diseases were far more widespread and deadly than they are today. In Cincinnati, for example, the three biggest killers in 1886 were tuberculosis, pneumonia, and bronchitis, and 31 percent of all deaths were lung-related. Data from Germany and England linking smoke to lung disease began filtering into the United States soon after. A 1905 German study found more acute pulmonary illness in smoky areas than elsewhere, and also discovered that animals infected with tuberculosis died more quickly in smoky air. An important 1914 Pittsburgh study confirmed the link between smoke and pneumonia deaths, adding that pneumonia not only killed the poor but "many of our most useful business men . . . on whom most has been spent on education."

By this time, there were other economic arguments for reducing smoke. Various attempts were made in the first decade of the twentieth century to calculate the cost of smoke, including losses due to the corrosion of stone and iron, the more frequent laundering of clothes, the extra cleaning of buildings, furniture, paintings, windows, and carpets, the premature replacement of wallpaper and draperies (generally replaced with dark colors to hide future damage), the extra energy spent on lighting, and the loss of retail merchandise stained by soot. Annual costs of this so-called soot tax were estimated to be $500 million a year, not counting health costs. While smoke abatement efforts would have been expensive, the soot tax was probably costing cities even more.

Who was causing all this damage? Residential use certainly contributed, but the vast majority of smoke seems to have come from other sources. Not surprisingly, the attention of activists and early regulators fell on industrial facilities, large commercial buildings, and railroads. Civic groups pressured railroads to stop burning coal in the cities altogether and switch to electric locomotives. After years of resistance, many did. Others were not asked to give up coal but merely to burn it more efficiently. Smoke, it was argued, was largely unburnt carbon, and if fires were hot enough and had enough oxygen, they could burn "smokelessly," that is, without visible emissions. In fact, it was estimated that 8 percent of the nation's coal, or $40 million a year, was wasted in the form of smoke. If coal users replaced their furnaces with better models, or retrofitted them to make them burn hotter, or made simple operational adjustments, the smoke problem would literally disappear.

By this time, an important shift was taking place in the smoke abatement movement. It was no longer in the hands of the women's or other civic clubs, or of the medical profession; it was in the hands of engineers. Smoke abatement was less and less a question of aesthetics, morality, civic pride, or even health, but a matter of simple efficiency—of not wasting coal. Around the nation, engineers took on the job of city smoke inspectors, and while some were tougher than others, all shared the belief that coal smoke could be engineered away, often just by tinkering with the furnace. Blue skies were right around the corner—if people would only apply proper engineering principles. These regulators were experts in the burning of coal, and their goal was to spread their expertise. The bigger question—whether something other than coal could be burned—was simply not on the agenda. The issue was smoke, not coal. As a result, those fighting the hardest against tougher pollution laws were not the coal producers but the major coal consumers.

In fairness to these regulators, few people then could imagine a world not heavily dependent on coal. In 1910, natural gas and oil still represented less than 10 percent of the nation's energy supplies, slightly less than wood. Even that meager contribution was assumed to be short-lived by many; Pennsylvania's oil and gas fields had already been largely depleted. In 1925, it was estimated that oil supplies would run out in fourteen years, and gas supplies were also considered too "ephemeral" to matter.

What impact did the clubs, the doctors, the engineers, and the new laws have on the air? Some cities claimed substantial improvements. In 1916, Pittsburgh officials reported that they had reduced the city's smoke by 46 percent just four years after adopting a new smoke ordinance. Progress was not this great in all cities, and none of these estimates were very scientific, but there was clearly a trend toward cleaner air in the cities that had taken smoke seriously. Unfortunately, this progress had largely depended on industries making capital investments in improved technology, continuing to operate their furnaces with care and attention, and burning costlier and cleaner anthracite or the better grades of bituminous. When industry's focus shifted to other priorities, such as winning the war to end all wars, the skies darkened again.

THE GERMAN EMPIRE, as John Maynard Keynes would later say, "was built more truly on coal and iron, than on blood and iron." The Germans were still third in world coal production and industrial strength, after the United States and Britain, and World War I would be very much a coal-driven war.* Coal didn't merely provide the fuel and steel critical to warfare—coal's hydrocarbons were the raw material for a substantial portion of the chemical industry. The most deadly explosives on the battlefield were mainly coal products, as were many of the new medicines and disinfectants in the hospitals where the soldiers were treated.

---

*It was common to attribute the relative industrial and military weakness of other European nations, and especially of France, to their relative lack of coal. Harvard professor of plant morphology E. C. Jeffrey opined in 1925 that "the so-called decadence of certain of the European races is clearly not due to any real degeneracy, but rather to poverty of resources in coal. This is notably true of the Latin races and is only less obviously the case for those Nordic nations which are without the indispensable mineral of our modern civilization." (He also offhandedly dismissed the troubles of Ireland on the nation's lack of coal rather than on oppression by the British.)

U.S. coal production soared to meet wartime demand. Smoke abatement concerns fell by the wayside because, as the federal government explained, "war meant smoke." The war also meant the most acute fuel shortage in U.S. history. During the severe winter of 1917–1918, some schools were closed for weeks, and Washington adopted fuel savings measures denounced as "worthy of a Bolshevik Government." Across the nation, coal was rationed to two-thirds the usual supply, and many uses of electricity—like most outdoor lighting and taking an elevator below the third floor—were forbidden. Daylight savings time was introduced to make better use of the sun's light.

When the war ended, the nation entered a period of unprecedented labor unrest. Like other workers, miners had come out of the war years with expectations of a more just world, but it would be a very long time before they saw it. Since their beginnings, labor relations in America's coal fields have been bloodier than in any other American industry, and bloodier than in Britain's coal fields. Strikes were often marked by a beating here, a shooting there, and sometimes by outright massacres.* In 1921, though, one labor conflict took a surprising twist that reflected the new postwar era.

---

*Some of these massacres were at the hands of the National Guard. In 1877, twenty-six unarmed miners were shot in Pittsburgh. In 1887, at least nineteen unarmed miners were shot in Lattimer, Pennsylvania, many in the back. In 1914, the Colorado National Guard, which included many former mine guards, attacked a tent colony where miners and their families, evicted during a strike, had been living through the long cold winter. They shot at and set fire to the tents, killing three miners as well as two women and eleven children who had huddled together in a tent for safety.

The labor drama in Logan and Mingo counties in West Virginia began the same way other coal field struggles had begun: with a strike, the mass evictions of miners and their families from company-owned houses, and the formation of a tent colony where the families lived for months. This time, though, with many of them recently returned from the trenches of Europe, the miners formed a well-armed and disciplined army some 6,000 strong and went on the march. War correspondents who had previously reported from European front lines now sent back alarmed dispatches from West Virginia. At the last minute, the United States Army rushed in, and a major bloodbath was avoided. Tens of thousands of homeless, hopeless miners went back to work. As the nation turned to other fuels, their industry slid into a long decline, bypassing the prosperous twenties altogether and getting a lengthy head start on the miseries of the Great Depression.

Roosevelt's 1933 New Deal and pro-labor legislation brought new hope to the miners, and sparked the resurgence of the UMW. The man who came to personify the American coal industry for decades was now at the union's helm: John L. Lewis, a burly former miner as famous for his bushy eyebrows as for his stirring rhetoric. Following the most successful organizing drive the nation had ever witnessed, the UMW was once again the nation's strongest union. From this position Lewis and the UMW won contract terms finally improving miners' lives, and then went on to change the course of the entire American labor movement.

The UMW was always an industrial union, which meant it covered all mine workers, regardless of skill level. Most other American unions, banded together under the American Federation of Labor (A.F. of L.), were organized along craft lines; they covered, for example, only carpenters, regardless of their industry, and excluded unskilled workers. Lewis believed that only industrial unions like the UMW could meet the needs of the growing mass-production industries like steel and auto, which either employed unskilled labor or too many skills to organize by craft. In 1935, he made a historic break with the A.F. of L., launching what became the Congress of Industrial Organization, or CIO, with the mighty UMW at its core.

Within months, workers in other industries were staging spontaneous sit-down strikes, forming new industrial unions, and flocking to the CIO by the millions; never would the labor movement cover a greater share of American workers than during this surge of membership. Whereas in 1902 John Mitchell was the nation's model of a conciliatory labor leader, by 1937 John L. Lewis was the model of labor militancy. More than that, he was the leader of a revived American labor movement that was patterned after the organizing philosophy of the coal mines.

For a time, Lewis filled stadiums with cheering supporters wherever he went, and was seen as the nation's second most powerful person after Roosevelt; but World War II changed that. By 1949, Lewis had become one of the nation's most hated public figures for leading the coal miners out on strike even during the war. Coal miners had become among the best paid industrial workers in the nation, but the support they once had from the American public had turned to anger.

One place Lewis remained popular was within the oil industry. With every coal strike, more of the U.S. energy market went to oil. One Venezuelan oil producer pondered erecting a statue of Lewis in Caracas "to honor him as one of the greatest benefactors and heroes to the Venezuelan oil industry." In fairness to Lewis and the miners, though, coal would eventually have lost its market share anyway. Coal is by its nature less compatible with modern technologies and modern consumer demands than oil or natural gas. Its solid form was an advantage back in the days of wagons and wooden ships, before pipelines and tankers; but in the modern world solidity just meant extra labor for those extracting it, moving it, and burning it.

In the affluent years after the war, as natural gas and fuel oil became more available, people turned away from coal, scrapping their big stoves, dusty furnaces, and ash cans forever. With no more domestic market, anthracite largely disappeared as a major industry, leaving the anthracite counties of Pennsylvania economically devastated.

And, of course, there was the rise of the automobile. Back in 1924, one coal industry analyst looked forward to a day when he could run his car for a month on a "couple of shovelfuls of coal," but that day never came; the transportation system, even the trains, ran on oil. During World War II, coal still accounted for about half of U.S. energy consumption, but then it began to fall, both in absolute terms and as a percentage of U.S. energy use;

by 1955, coal had nose-dived to less than 29 percent of the nation's energy use.

For nearly a century, King Coal was arguably the most visible and controversial industry in the nation. In this center-stage role it left an indelible mark on many core features of American politics, including its centralized business structure, its deep anticorporate sentiments, its antipollution movement, its approach to environmental regulation, its labor relations, and its union movement. Now the industry was entering a period of diminished influence in every respect. But while coal was down, it was not out. It was merely taking on a new role in American life, this one largely played behind the scenes.

seven
....

Invisible Power

Aʙᴏᴜᴛ ᴀɴ ʜᴏᴜʀ's ᴅʀɪᴠᴇ ɴᴏʀᴛʜ of my home in Minnesota three enormous pillars of fire rage night and day. Each one is about ten stories high, and each is contained like a caged beast inside gigantic boilers at a power plant called Sherco, the flagship facility of my local power company. Driving up to Sherco one June morning, I could see from miles away its pair of smoke stacks looming over the landscape, each reaching higher than the Washington Monument. As I pulled up to the front gate for a scheduled tour, I couldn't help but feel slightly intimidated by the sheer physical power before me—power both wonderful and, given its global impact, slightly sinister. The gate of this bastion of modern, large-scale coal power was guarded by a smiling, grandmotherly security guard who took my name and sent me in with a friendly wave.

I was greeted at the plant doors by two genial and enthusiastic tour guides, Jack and Glenn. Both semiretired engineers, they had spent their careers working at power plants, including many years at Sherco. They did not flinch when I told them I was writing a book about coal and its pollutants, or when I mentioned my background enforcing air pollution laws. They were deeply proud of their facility, and honestly pleased to show it off. As we waited in a classroom for the rest of the tour group, my guides pointed out framed pictures of the power plant and handed me jars of pulverized coal, ash, and slag to admire, as if they were showing me pictures of their grandchildren.

The rest of the tour group arrived: a score of high-school students and their teacher. Jack and Glenn worked hard to convey the sheer magnitude of what we were about to see. They explained how the three boilers at this one plant make enough electricity to power over 2 million homes, an amount that dwarfs the output of the nuclear plant that sits not far away. They explained how the construction of the newest boiler in the 1980s cost a billion dollars, making it then the costliest construction project in state history. The slouched and sleepy students were unimpressed; they showed no appreciation for the Promethean effort and enormous social investment that Sherco represents, nor did they appear to have any awareness of the environmental and climate threat that, sadly, even friendly power plants like this one pose.

We all donned hard hats and radio headsets through which our guides could speak to us over the din of the equipment, and then began our tour. The plant is several stories high, and because in some places the floors are made of transparent steel grating, you can see the piping and equipment on floors far above and far below. A girl with vertigo panicked when she looked down and had to quit the tour. Everything was so mechanized that we saw only a handful of people. First, we saw the massive pulverizers that grind coal into a black powder finer than flour, and then the equipment that blasts the coal powder and heated air into the three boilers; next, we went to see a boiler, one of the bellies of this three-bellied beast with the massive appetite. Together, Sherco's boilers consume about 6.5 million tons of coal a year. That is about three-fourths of what the entire nation consumed in 1850, a time when coal was already transforming the American economy.

Jack explained that each boiler is lined with hundreds of miles of steel piping. In these pipes water becomes superheated steam with enough pressure to make the nearby electric generators spin at twice the speed of sound. In time, coal residue builds up within the boilers, not unlike the clinkers that used to grow in coal stoves and furnaces; but in these boilers the residue can grow to chunks "the size of a minivan." These superclinkers have to be blown loose with dynamite by explosives experts who make their livings climbing into cooled-off boilers for this purpose.

At the boiler, there was at first not much to see. Rising up through several floors of the plant, the boiler up close was just a warm, vibrating metal wall. It was hard to believe that on the other side was a 3,000-degree Fahrenheit fireball some forty-five feet across and ten stories high devouring up to five hundred tons of powdered coal each hour. But then Jack nonchalantly opened a tiny door in the boiler's side, just a crack. We shielded our eyes as a blinding white light poured out, like sunshine held captive underground for millions of years and finally set free.

Thanks to fires like these, the recent history of coal fires in the United States, and in most developed nations, has been marked by increasing size and decreasing visibility. From the consumer's perspective, coal has virtually disappeared—its sooty black chunks magically transformed into squeaky-clean electrons. Now that nine out of every ten tons of the nation's coal vanishes into power plants, many Americans can harbor the illusion that coal is no longer a major energy source or a big environmental threat, even while the nation burns more of it than ever.

Coal use in the United States can't remain invisible for much longer, though. It is increasingly under attack on environmental grounds, and objections are coming both from home and abroad. Even after decades of regulation, an astounding proportion of the most serious pollution problems in the United States are still caused by coal, and the threat of global climate change—something U.S. laws have yet to touch—is a matter of increasing international urgency. Watching U.S. investment in coal continue to rise despite the simultaneous growth of these profound environmental concerns is a little like watching two trains, each with tremendous momentum, race toward each other on the same track. It's too soon to say exactly when the crash will occur, but a loud, messy, and painful collision is inevitable.

FOLLOWING COAL'S EMISSIONS through nature is an ongoing scientific odyssey. Already, it has forced us to increase our understanding of how the wind blows, how the rain falls, how light scatters, and how chemicals cycle through air and soil and water. It has also taught us things about our own lungs and hearts and brains, and about the subtle and sometimes not-so-subtle ways that things we release into the world can come back to haunt us. All in all, tracking the smoke has given us ample proof that we're intimately linked, through our economic decisions and through large-scale natural processes, to the broader web of life on Earth and to each other.

The first great wave of environmental sentiment swept through the nation in 1970, marked by celebration of the first Earth Day. In the wake of this new awareness came the formation of the Environmental Protection Agency (EPA) and passage of a string of new laws, the most ambitious and far-reaching of which was the 1970 Clean Air Act. The act established the straightforward requirement that the nation's air be cleaned up to healthy levels by 1975. To a nation that had just landed men on the Moon, such a lofty goal must have seemed entirely possible. Three decades later, it looks hopelessly naive, a reflection of how little we knew about air pollution, health, and environmental politics.

When the Clean Air Act was passed, the nation's primary concern was the effect of urban air pollution levels on health. The potentially deadly nature of urban smoke had been demonstrated some years earlier during London's historic "Black Fog" of December 5–9, 1952. A temperature inversion trapped the city's smoke close to the ground. On the first day it was still a white fog, but so extraordinarily dense that cars and buses moved slower than a walk, and the opera had to be cancelled when fog seeped into the theater and made it impossible for the singers to see the conductor. By the last day, the fog had turned black, visibility was limited to a mere eleven inches, and the hospitals were full of Londoners perishing from the smoke. Many of the 4,000 or so people killed by this episode never made it to the hospital but died on the streets; fifty bodies were removed from one small city park. In 1956, after nearly seven hundred years of complaints about the coal smoke in London, Parliament finally banned the burning of soft coal in the central city, and the air immediately improved.

The primary culprit in the deaths was a gas called sulfur dioxide. Sulfur dioxide, or SO_2, is not some exotic chemical contrived in a laboratory. Nature spews it out through volcanic eruptions, but we spew out more with our coal fires, our own slow-motion, domesticated little volcanoes. This old-fashioned and surprisingly dangerous pollutant results, quite literally, from mixing fire and brimstone, or, in other words, from burning the sulfur that contaminates coal.

In 1970, total SO_2 emissions in the United States were reaching an all-time high. The greatest sources by far were the coal-fired power plants, which doubled their SO_2 emissions every decade between 1940 and 1970. Despite these increasing SO_2 emissions, in most city centers the air probably held much less SO_2 than it had earlier in the century because taller smoke stacks were spewing it farther. Some of the Clean Air Act regulations promoted the continuation of this trend. A common saying among environmental regulators was "the solution to pollution is dilution." Of course, this is only true if your fires are small enough and your planet is big enough, and neither turned out to be the case. In time, the dirtiest air got cleaner but downwind some of the cleanest air got dirtier. Health concerns persisted, but they were soon overshadowed by other issues, like the mystery of the missing fish.

In the late 1960s, scientists were perplexed over the disappearance of fish that had formerly thrived in certain pristine lakes in southern Sweden and Norway and the Adirondack Mountains of New York. The lakes looked more beautiful than ever because the microorganisms that used to cloud the water were also gone. There were other mysteries, too: High-elevation stands of trees were dying of unknown causes, and, oddly enough, in parts of Sweden people were surprised to find their hair turning green. Researchers finally linked all these problems back to acid-forming pollution, and particularly to SO_2 emissions from the enormous coal fires in Britain and Germany (causing pollution in Scandinavia) and in the industrial Midwest (causing pollution in the northeastern United States and Canada).

Acid rain not only raises the acidity of distant lakes and streams, which directly harms fish, but as it passes through the soils it also leaches out toxic minerals like aluminum and mercury, causing more damage to the ecosystems. By the 1970s the acid had seeped into thousands of shallow wells in southern Sweden, corroding the copper pipes and contaminating the tap water with copper sulfate. The fair-haired Swedes who washed in the contaminated water found that it tinted their hair green, sometimes as "green as a birch in spring," as a Swedish researcher described it in 1981. It was an ironic choice of words, because the same pollutants were causing certain trees to lose their green needles and turn brown.

As the evidence linking acid rain to coal poured in, the response of the electric and coal industries was to deny the link, to question the motives of those investigating the connection, and always to call for more research. In October 1980, the head of the National Coal Association dismissed acid rain as "a campaign of misleading publicity which seems designed to gain public support for new legislative and regulatory measures." He need not have worried about new environmental laws, though, because the next month Ronald Reagan was elected president. There would be no significant new effort to address acid rain for ten more years. Existing EPA regulations were forcing a gradual decline in SO_2 emissions during this time, but not fast enough. Finally, in 1990, Congress adopted an Acid Rain Program requiring power plants to cut their SO_2 emissions nearly in half by 2010.

Since 1990, when the acid rain issue disappeared from public view, there have been two big surprises. The first is that the SO_2 cuts have cost so much less than anyone thought—about twenty times less than some early industry estimates, and even less than environmentalists had predicted. SO_2 reductions have been a bargain thanks to cheaper-than-expected low-sulfur coal from the western United States, improving technology, and flexible regulations that allow the power industry to choose the cheapest means of compliance.

The second big surprise is less welcome: There is now evidence that the long-delayed and hotly debated cuts required by the 1990 law may not be enough. While acid rain has been reduced, average rainfall in many parts of the United States is still occasionally ten times more acidic than normal. After decades of being showered with acids, some areas have had the neutralizing minerals found naturally in the soils washed out, leaving them substantially more sensitive to newly arrived acids. The Canadian government has found that even after full implementation of the U.S. Acid Rain Program, thousands of Canadian lakes in an area the size of France and the United Kingdom combined will continue to acidify. Environmentalists point to studies showing acid-caused damage ranging from declining salmon stocks in Nova Scotia to the loss of trout in Virginia streams to the decline of red spruce and sugar maple in the Northeast, and are calling for additional SO_2 reductions of up to 80 percent beyond the 1990 requirements. Research suggests that even with cuts this deep, it will take up to a quarter of a century for some ecosystems to shake off the acidity accumulated over decades of pollution.

Even before the evidence of the continuing acid rain problems emerged, coal-burning was being increasingly blamed for a surprising proportion of our other long-standing pollution problems. For example, we know now that SO_2, invisible when it comes out of the smokestack, is the primary reason people can't see great distances in the eastern United States. On an average day in the East, you can see about fourteen miles. If not for human-made air pollution, you could see from forty-five to ninety miles. The main cause of the vista-destroying haze that covers virtually everything east of the Mississippi are tiny sulfate particles that scatter sunlight; the particles are formed from SO_2 emissions, which come mostly from coal fires. In the West, even though SO_2 emissions are much less, there is still enough coal pollution to obscure views: The EPA found that a significant part of the haze problem at the Grand Canyon has been due to SO_2 emissions from a coal plant seventy-five miles away. One economic study found that the Acid Rain Program's improvement of visibility—a benefit barely considered when the law was passed—is alone worth the substantial cost of pollution controls, quite apart from the many other environmental and health benefits.

Of course, vistas are also impaired by what we call "smog." Smog is mainly ozone—a gas we have too little of up in the stratosphere, where it shields us from radiation, but too much of down here. Ozone is a big problem both in rural areas, where it harms the growth of forests and crops, and in urban areas. There has been some gradual improvement in the nation's average ozone levels over the last twenty years, but in some areas the problem is getting worse, and over 81 million Americans still live in areas that fail to meet EPA's health-based standard. Studies in the northeastern United States and Canada have found that from 10 percent to 20 percent of summertime lung-related hospital admissions are linked to ozone, which can trigger asthma attacks, increase the risk of respiratory infections, and irreversibly change lung structure. Children are most at risk.

Ozone does not come out of smoke stacks. Rather, it forms when gases called nitrogen oxides (NO_x) combine with other air pollutants in the presence of sunshine and heat. Although smog is mainly blamed on road traffic, coal contributes nearly a quarter of the nation's NO_x emissions, which is more than all cars, vans, and sport utility vehicles combined. NO_x emissions are also a secondary cause of acid rain, and, because nitrogen is a fertilizer, they contribute to excessive algae growth and the depletion of oxygen in coastal waters such as the Chesapeake Bay.

And then there is the slippery problem of mercury. Many thousands of lakes in the United States contain fish so tainted with mercury that pregnant women and children are warned not to eat them because mercury can damage the developing brain. Most mercury in lakes rains down from the air, and perhaps as much as a third of mercury emissions to the air comes from coal plants, making them the largest source. The public health threat posed by mercury is not trivial: A recent report by the National Academy of Sciences warns that 60,000 babies born in the United States *each year* may have been exposed to enough mercury in utero to cause poor school performance later in life. Once mercury is introduced into the environment, it is impossible to clean up because it keeps reevaporating and raining down again indefinitely, ping-ponging its way mercurially around the world, posing new risks wherever it lands.*

We may think that because coal smoke is dispersed so far and wide we no longer need to worry about concentrations high enough to kill us, but unfortunately we do need to worry. In fact, there's now alarming evidence that coal burning is killing not just a few of us but many thousands every year in the United States alone. This is a stunning finding in the field of environ-

---

*Although mercury is the air toxic of most concern, coal plants also emit dozens of other highly toxic chemicals, but in relatively low amounts. In particular, the EPA has expressed concern over emissions of the carcinogens arsenic and dioxin from coal plants, and has announced plans to regulate not only mercury but also other toxics from power plants.

mental regulation, where it's exceedingly rare to find huge numbers of deaths linked to one pollutant or to one industry.

Throughout the nation, devices that monitor the levels of SO_2 in the air suggest that SO_2 no longer threatens our health. Almost everywhere, SO_2 has been reduced to levels the EPA says pose no health threat. The problem, though, is that by the time we measure for SO_2, much of the gas has already performed a sort of chemical sleight-of-hand, transforming itself into particulate form, and in so doing, becoming much more deadly. Both SO_2 and nitrogen oxides—the pollutants already linked to acid rain, loss of visibility, smog, and the over-fertilization of waterways—undergo this dangerous change.

Particulates, tiny things that float through the air, can consist of anything—acids, toxic metals, grain dust, dirt—and they can be solid or liquid. All particulates have been regulated the same way, partly because it's so hard to tell them apart, and partly because their size alone makes them dangerous. The smallest ones can sneak past your body's defenses, below the sneeze threshold, past the cilia of your nose and bronchial tubes, and all the way into the deepest and most sensitive recesses of your lungs. It took us a while to realize that the very smallest particulates are the deadliest and that they come mainly from burning fossil fuels, especially coal and diesel fuel.

Despite the suggestion by an early twentieth-century coal merchant that soot in the lungs is healthy because it helps filter the air as it passes into the blood, it isn't good to have particles deep in your lungs; on the contrary, it increases your chances of dying from lung ailments (emphysema, bronchitis, asthma) and heart conditions. How many deaths we are talking about is hard to estimate. Epidemiological studies have come a long way since London draper John Graunt began looking at death rates in the seventeenth century, but they still involve major uncertainties. The evidence strongly suggests, though, that those killed by particulates number in the thousands, and quite possibly the tens of thousands, each year. Researchers at the Harvard School of Public Health have estimated that from 60,000 to 70,000 deaths per year are linked to particulates from all sources.

A study commissioned by an environmental group and conducted by consultants who routinely perform research for the EPA focused on the overall health impact of particulates emitted from the nation's power plants. It estimated that power plant emissions are killing over 30,000 people a year, and the vast majority of those emissions are from coal plants. In addition, these pollutants were found to cause tens of thousands of hospitalizations, hundreds of thousands of asthma attacks, and millions of lost work days yearly.

If these estimates are even close to accurate, then coal burning is a public health threat of the first order. It would mean that coal kills nearly as many people per year as traffic accidents (42,000 in the year 2000), and causes more deaths than homicide (16,000) or AIDS (14,000). And burning coal in the United States takes far more lives than digging it takes: Black lung kills maybe fourteen hundred coal workers each year, and coal mining accidents kill around forty miners in a bad year. The EPA is now tackling particulates with new regulations after surviving a five-year battle during which industry groups, including power companies, brought the regulations all the way to the Supreme Court. It will, however, be many years before the new standard is translated into cleaner air.

The coal industry looks at reports linking coal to health threats with a deep sense of injustice; it feels that the nation has been insufficiently grateful for all the cheap energy it has provided and for the significant strides it has made in pollution control. One coal company's advertising slogan hails coal power as "Essential, Affordable, and Increasingly Clean," and the industry stresses that although power plants now burn three times the coal they burned in 1970 when the Clean Air Act was passed, concentrations of SO_2 in the air, and to a lesser extent NO_x, have been declining. Why, after all this progress, is coal now the target of so many legal proceedings and headlines suggesting the problem is getting worse?

The answer is probably that even though the levels of these life-threatening pollutants are indeed lower now than before, we now know much more about the impact of what's still there; the high remaining risk is more newsworthy than the even higher risk we faced in our more ignorant past. Coal burning in the United States may be less deadly than in years past, but nobody's quite sure what to do with this good news. Those concerned about today's heavy death toll are not much impressed when they hear that it probably used to be heavier, and those drawing attention to pollution reductions aren't ready to accept that the emissions are still quite deadly. A slogan like "Coal—Essential, Affordable, and Killing Thousands Fewer Americans

Than It Used To" may be closer to the truth, but nobody is going to base an ad campaign on it.

The billions of dollars spent on SO_2 reduction in the last few years have mainly been spent in the name of lakes and trees and fish, under the acid rain regulations. Nature has indeed benefited, but the cuts have probably also saved thousands of lives, mainly of people who have no idea how much they owe to an effort to protect distant ecosystems, or how intimately their fate and that of the natural world are linked.

THERE MAY BE NO POLLUTANT in all of history that people have worked harder to defeat than sulfur dioxide. In the United States alone the battle against it has absorbed years of effort and billions of dollars. Although it is nowhere near won, it has already utterly transformed the coal industry. It has also created deep political fissures between states and regions, as local fortunes rise and fall and as states struggle to decide how much they are willing to sacrifice to fight this invisible foe and to stop killing their downwind neighbors.

Coal provides just over half the electricity for the United States, with huge and politically important regional variations. Many states, especially in the coal-producing regions, get virtually all their electricity from coal; others, especially on the West Coast and in New England, burn next to none. They rely instead on the three power sources that make up almost all the other half of the nation's electricity—hydroelectric dams, nuclear power, and natural gas—though sometimes they also import coal-fired electricity from other states. However, the number of people a state's coal plants actually sicken or kill (and the amount of acid rain and lost visibility they cause) depends not just on how much coal they burn but on the kind of coal they use and how they burn it.

People who run coal-fired power plants can cut their SO_2 emissions in two ways: They can scrub, or they can switch. "Scrubbers" are costly pollution control devices that spray a substance through the boiler's exhaust; the sulfur in the exhaust chemically bonds with the spray and can then be collected. Most scrubbers use a liquid spray and create spectacular amounts of fairly toxic sludge that usually ends up in a landfill nearby. Some scrubbers spray in a powder instead; but then the powder has to be filtered out of the exhaust by using, in essence, a fabulously large vacuum cleaner. At the Sherco plant in Minnesota, they vacuum their emissions with a device called a "baghouse" nearly the length of a football field and containing over 18,000 vacuum bags, each about as long as a school bus, the contents of which must be continually emptied and disposed of. Scrubbers cost hundreds of millions of dollars, but can reduce SO_2 emissions by anywhere from 70 to 90 percent.

It's often cheaper and a lot easier, though, just to switch to a kind of coal containing less sulfur. This option has caused some painful changes to the traditional U.S. coal industry by wrenching much of it away from the high-sulfur coal fields of the East and moving it to the low-sulfur coal fields of the West. Western coal has always been easier to dig because it lies in thick seams near the surface; but it is younger and generally packs less of an energy punch than the older eastern coals. In 1970, before environmental laws made sulfur content so important, only a tiny share of U.S. coal came from west of the Mississippi. Today, more than half of it does, and the growing western low-sulfur coal fields are continuing to drain business away from the suffering high-sulfur eastern fields. Wyoming, with its vast surface mines, is now the nation's top coal-producing state. One result of the shift is that the coal industry can no longer view the environmental movement solely as a threat; the western coal fields, at least, owe much of their growth to environmental laws, and could benefit even more from stricter SO_2 limits.

The shift to the western United States has transformed this ancient industry in other ways, too. Even though coal fueled the rise of the machine in virtually every sector of the economy, the extraction of coal (as opposed to mine drainage or coal transportation) long depended far more on manual labor than on machines. Today, though, nearly two-thirds of American coal comes from surface mines, where it has been scooped up by some of the world's most gargantuan machines. This mechanized mass production has helped make the direct cost of coal incredibly cheap by any measure.

The mechanization of mining has also devastated the work force. Because only the largest mines can afford to mechanize, the smaller mines that characterized the soft-coal industry for so much of the twentieth century have finally been driven out. More than three-quarters of the coal mines operating in the

United States in 1976 have closed, and the current work force of 72,000 coal miners is less than a third of what it was a quarter century ago. The once mighty UMW now represents a mere 20,000 miners, who produce less than a fifth of American coal.

Although most eastern coal is high in sulfur, a few pockets of low-sulfur coal are found in Appalachia. In the desperate effort to reach them, people literally move mountains, sacrificing the land in the name of protecting the air. In southwestern West Virginia—in the same counties where an impromptu army of thousands of evicted coal miners marched in 1921—perhaps one fifth of the mountains have already had their top few hundred feet blasted and scraped away by a relatively new mining technique aptly called "mountaintop removal."

The rubble that was once the mountain peak is being dumped in the valleys, accelerating by hundreds of millions of years the erosion processes that have already brought the Appalachians down from their formerly Himalayan heights. Reclamation can eventually add a layer of vegetation to the new landscape, and the industry points out that leveling the land makes it more suitable for commercial development—such as the golf course that has been built on one reclaimed site. Still, the practice has created one of the most bitter struggles this coal state has seen in years, pitting those who want to hang on to dwindling coal jobs against those who hate to see their mountains and valleys forever altered.

There are, however, many coal plants that have still neither scrubbed nor switched, infuriating the states downwind of them. The most notable are the many old plants in the historic eastern coal regions along the Ohio River valley and Appalachia. Federal regulations requiring scrubbers were adopted many years ago, but they applied only to future plants; it was assumed the old ones would close down in a few years. Instead, the owners of these plants, which can run more cheaply than cleaner plants, have replaced one piece of them at a time in a way that renders the plants virtually immortal. Extending the old plants' lives without installing scrubbers is illegal, according to the EPA, which under the Clinton administration launched one of the biggest enforcement cases in its history over changes at fifty-one old power plants. The Bush administration, though, has since questioned the enforcement action, raising concerns that it will either be dropped or narrowed.

For years, downwind states in the Northeast, especially New York and the New England states, have loudly complained about the pollution the still-uncontrolled plants send their way on the prevailing winds, which can carry the pollution hundreds of miles. Since they receive substantial amounts of deadly particulates and acid-rain-forming pollution from these upwind plants, they truly have something to complain about. On the other hand, the heavily coal-dependent states are paying a hefty price of their own for their SO_2 emissions. The greatest health risk appears to be only twenty miles or so downwind of the plants, despite their tall stacks. It appears these coal states have paid for their lower electricity rates (and temporarily lower unemployment rates among their miners) with higher death rates. The coal deposits that have so hugely influenced their patterns of growth throughout their history continue to influence their patterns of death today.

After years of largely piecemeal regulation of coal plants, Congress is now looking at bills that would require huge cuts not just in SO_2 but in NO_x and mercury as well. Compliance would not be easy. For the most part, different technologies are needed for different pollutants, none of the approaches are cheap, and all have serious environmental down-sides—among them, the disposal of the resulting sludge. Still, if the politicians could find a way to require substantial pollution cuts, it appears that the engineers could find a way to accomplish them. Moreover, the government estimates that even with much stricter and costlier SO_2, NO_x, and mercury controls, burning coal would still be economically worthwhile.

However, some of the legislation before Congress also addresses a fourth pollutant, one that poses environmental threats on a vast new scale. That pollutant is carbon dioxide, the main cause of global climate change, and it cannot be controlled with existing technology. If regulated even modestly, it would change the future of coal use entirely. To the coal industry, this is the straw that threatens to break the camel's back. Although industry leaders have made that claim before (indeed, many times), this time, they might be right.

IT'S HARD TO THINK that a gas as friendly as carbon dioxide can be a pollutant. Carbon dioxide puts the bubbles in your soda pop and the holes in your bread. It extinguishes fires. Frozen, it's dried ice. It isn't noxious, or caustic, and it doesn't damage lungs, poison ecosystems, or destroy vistas. The coal industry sometimes calls it "harmless-to-health carbon dioxide," and for the most part they're right. Ironically, the people who throughout history have had the most to fear from carbon dioxide are coal miners; CO_2 is the "choke damp" that has for centuries smothered miners when it builds up in a mine. At normal atmospheric levels, though, CO_2 poses no direct threat.

In fact, CO_2 is essential to life on earth. We earthlings are carbon-based life forms, and so are the plants around us. Carbon dioxide is the gaseous part of the carbon cycle you probably studied in science class. You and other animals exhale CO_2 with every breath, and plants draw it in to carry out photosynthesis, turning the carbon in the air into the carbon of their bodies. The most delicate blades of grass and the most massive trees have been formed, in a very real way, out of thin air. Plants ultimately release most of that carbon dioxide again when they decay. Every year, when spring comes to the Northern Hemisphere (where most of the earth's land mass is), the rejuvenated plant life draws in the CO_2, and global levels of the gas in the ambient air decrease measurably. In the autumn, when the leaves fall and decay, the CO_2 returns to the atmosphere. The biosphere, in effect, breathes in huge synchronized annual breaths.

Carbon dioxide plays another critical role in the preservation of life. It is the earth's most important greenhouse gas, so called because it traps heat in the same way the glass of a greenhouse does. It makes up just a tiny bit of our atmosphere, but that little bit goes a long way: Without CO_2 here keeping the warmth of the sun from bouncing back into space, the planet might be nothing more than a lifeless frozen ball. It would be a mistake, though, to view CO_2 as purely benign. As with other major forces of nature—water, sunshine, and wind—whether CO_2 is good or bad, life-giving or devastating, depends largely on whether it is in balance with the other forces. On Earth today, that balance is being threatened.

The very essence of coal is carbon that was taken out of circulation over millions of years; burning coal suddenly puts that carbon back into play, making the world a bit more as it was when the coal was originally formed. It's almost as if the *lepidodendron* of old had found a way to use humanity to re-create the planetary conditions they once thrived in.

Since the dawn of the industrial revolution, we've burned enough fossil fuels to increase the amount of carbon dioxide in the air by about one-third, already bringing it to a level probably not seen in the last several million years. Projections show CO_2 could reach levels two or three times the preindustrial concentration, or higher, within the next century. And once CO_2 levels go up, it takes centuries for natural processes to bring them back down again. Not all of this build-up is from coal; oil and natural gas also contribute enormously.* Coal, however, creates substantially more CO_2 than the other fossil fuels for the amount of energy obtained (roughly twice the CO_2 that natural gas pro-

---

*In the United States, petroleum is the source of 43 percent of energy-related CO_2 (mainly from transportation) and natural gas is the source of 21 percent of emissions. Coal is the source of 36 percent of emissions, but it provides only 22 percent of total energy.

duces, and a third more than petroleum produces).* That is why environmentalists, regulators, and the coal industry all tend to see efforts to prevent climate change as the beginning of the end for coal.

There is plenty of evidence that the warming has already begun; the 1990s were the warmest decade since global record keeping began, around 1860. Indirect data from temperature proxies, like tree rings and corals and ice cores, indicate that the 1990s were probably the warmest decade of the last thousand years. Plants and animals are already beginning to shift their ranges in their efforts to follow the climate that suits them; permafrost is thawing, and on nearly every continent ancient glaciers are in rapid retreat. Vast pieces of the ice shelves on the fringes of Antarctica have collapsed into the sea with a speed that has stunned the scientists observing them. The limited measurements we have of the North Polar ice cap show it thinning by a startling 40 percent just since the 1950s, when submarines began measuring it from below.

The concern, though, is not over what has happened so far, but over the more dramatic warming ahead. The Intergovernmental Panel on Climate Change (IPCC), a group of over 2,000 scientists gathered by the United Nations to assess the complex science of climate change, predicts warming over the next cen-

---

*In electricity generating, the CO_2 advantage of natural gas over coal is even greater. The most efficient new gas plants produce less than a third of the CO_2 per megawatt that typical coal plants produce, and only 42 percent of what the most efficient new coal plants produce.

tury ranging from 1.4 to 5.8 degrees Celsius by 2100. Whether we end up at the higher or lower end of this range depends on which computer model turns out to be right and on how quickly greenhouse gases build up.

Even at the high end of that range, it's hard to be alarmed over such tiny numbers until you realize that these global averages can mask climatic changes of epic proportions. At the depth of the last ice age thousands of years ago, much of the land in the Northern Hemisphere was covered by an ice sheet about a mile high. The average global temperature at that time was only 5 or 6 degrees Celsius colder than today's. At the high end of the warming range, we are looking at a warming in only a century about as great as the one that melted that ice sheet, with more warming in the centuries ahead.

What will a warmer world look like? It will very likely include more frequent heat waves, droughts and forest fires, and, paradoxically, more flooding and landslides as greater evaporation comes down in more drenching rainstorms. Smog and water pollution problems will be compounded by the additional heat and floods. Slowly rising seas will threaten low-lying coastal areas around the planet—including many of our most productive ecosystems and most densely populated regions—with inundation and more destructive storm surges. Many plants and animals both on the land and in the oceans will no longer be able to survive where they are, and widespread ecosystem disruption will result.

Some species will shift their ranges to follow the climate conditions that suit them, as they have with natural climate changes of the past. Often, an area's transition from one ecosystem to another won't be pretty—imagine, for example, the great boreal forests dying off along their southern range and the withering of U.S. sugar maples as their range moves north out of the United States and into Canada. Moreover, because towns and roads and farms will be in their way, many species simply won't be able to migrate out of their fragmented habitats to a suitable new place and, as a result, some ecosystems will be irreversibly depleted. The threat of uncontrolled outbreaks of insects and invasive weed species will increase when existing balances between plants and animals, predators and prey, are disrupted. Global extinction rates will rise even higher, and many conservation efforts will have to shift from preserving existing ecosystems to trying to reestablish them in new places.

There will be benefits, of course, as with most great changes. Some species will thrive, there will be fewer deadly cold snaps, heating costs will decline, and in many places the growing season will lengthen. Crop and wood production may well increase over the next few decades in the mid-latitudes if warming remains moderate. This is thanks in part to more CO_2 in the air, which has a fertilizing effect that for many plants will help offset some of the other climate stresses.

Even so, we can be confident that the direct losses we face, quite apart from those represented by a depleted natural world, will be tremendous. We've made a huge investment in the existing climate and shorelines. They have largely determined where we've built our cities, our hydroelectric dams, and our water supply systems; how we've built our homes and what crops and

trees we've planted; whether our local economies run on farming or forestry or tourism—the list is endless. Much of that
investment is certain to be stranded in the decades ahead. Those
with enough resources will adapt, of course, but they'll have to
do so without the luxury of predictability as the climate of our
past becomes less of a guide to the future, and as variations from
year to year increase. Deciding which new crops and trees to
plant, how to restructure a city's water supply, or where to
rebuild after a flood will be a bigger gamble than ever.

In the poorer nations of the world, the costs promise to be
far greater. Food and water supplies are stretched critically thin
in many nations, and far more people are crowded into areas at
very high risk of flooding or desertification. Without the money
it takes to adapt, such nations could face grave food and water
shortages, the spread of tropical diseases across a greater range,
waves of refugees fleeing flooded or drought-ridden areas, and
the general destabilization of what are often fragile political
regimes. The world is much too small to imagine that these third
world problems will not generate global repercussions; climate
change promises to make a dangerous world more dangerous for
everybody.

The predictions above, frightening as they are, are the likely
result of a moderate to high warming that gradually unfolds over
a century. We may not be so lucky. In fact, fairly new evidence
about past climate changes tells us that a gradual change would
be uncharacteristic of our temperamental planet. Earth's climate
has been on generally good behavior for the past 10,000 years or

so, letting our species evolve a civilization. Historically, though, the climate has been inclined to move in sudden jolts and surges of a civilization-threatening magnitude.

A 2001 report ominously titled "Abrupt Climate Change: Inevitable Surprises," opens with this remarkable warning: "Recent scientific evidence shows that major and widespread climate changes have occurred with startling speed. For example, roughly half the north Atlantic warming since the last ice age was achieved in only a decade." These are not the words of an alarmist environmental group, but of the highly prestigious, buttoned-down, and generally understated National Academy of Sciences, an independent body chartered by Congress to advise the nation on matters of science. The academy goes on to warn that in the past such abrupt changes have been accompanied by severe and widespread floods and droughts, and that this new thinking about how quickly the climate can change is "little known and scarcely appreciated" by policymakers. Such large and unwelcome climatic jolts are particularly common when the climate is being forced to change most rapidly. In other words, in times exactly like these.*

---

*The mechanism for such abrupt changes could be the shut-down of the world's all-important ocean currents, such as the Gulf Stream, which distribute heat around the earth. Such a shut-down would cause "massive changes both in the ocean . . . and in the atmosphere," according to the report. Others have even theorized that the collapse of such currents would cause a North Atlantic cooling so severe that it could trigger the onset of a new ice age, though the National Academy report dismisses this possibility.

IN 1997, THE NATIONS of the world met in Kyoto, Japan, to try to make history: to reach a binding international treaty cutting greenhouse gas emissions. They hammered out an agreement under which the rich nations would reduce their greenhouse gas output to 5 percent below 1990 levels by the years 2008–2012. It was understood to be only a small first step in the right direction, and not nearly enough actually to stop the buildup in greenhouse gas levels in the atmosphere. Stopping the buildup would have meant immediately cutting global emissions by more than half, with deeper cuts to come. Even limiting the buildup of greenhouse gases to a level double the natural atmospheric level will eventually require measures far more ambitious than those proposed in Kyoto. Modest as it is, though, Kyoto still represents something potentially momentous: an agreement by the developed world to start the shift away from fossil fuels and toward an energy path radically different than the one we have been on for centuries.

Not everyone supports such a shift, of course. In the late 1980s and 1990s, U.S. energy and industry groups conducted a broad-scale public relations and lobbying campaign against any greenhouse gas limits. They were so effective that by the time Kyoto was negotiated, it was declared dead on arrival in the U.S. Senate, which would need to ratify it for the United States to be bound. The staunchest opposition to Kyoto or to any kind of limits on carbon dioxide has always come from the U.S. coal industry.

In response to climate change, the U.S. coal industry has done many of the things that late twentieth-century industries threatened by scientific findings typically do. They've sought out scientists with friendlier views, funded their research, and amplified their voices. They've accused their critics of having ulterior motives. They've launched public relations campaigns to influence public attitudes, including some highly misleading ones. They've relentlessly lobbied federal policymakers and gone from state to state to oppose all measures taking climate change seriously. One coal producer even sued environmental groups that had blamed coal for climate change, alleging that their environmental claims had defamed coal by casting it "in an unwholesome and unfavorable light."

At the high-profile vanguard of all these activities has long been a small, Colorado-based coal cooperative called Western Fuels Association. Western Fuels has also introduced a kind of rhetoric into the climate change debate not generally heard in political discourse, or at least not in recent centuries. Its president has argued that God gave coal to humanity to use in carrying out the biblical command to fill the Earth and subdue it, and he criticized governments for having "the arrogance to attempt to intervene in the normal, industrial evolution of mankind." This view strongly echoes nineteenth-century sentiments, but it comes with a modern twist. Not only is rising coal use seen as preordained but so are the rising carbon dioxide levels; they are part of the way humanity is reshaping the Earth to make it more hospitable for our expanding population.

Carbon dioxide is a plant fertilizer, commonly pumped into commercial greenhouses to promote plant growth. If more CO_2 benefits greenhouse plants, Western Fuels has argued, it will benefit nature as a whole.* The Earth's atmosphere "is deficient in carbon dioxide," and burning coal helps correct that. In fact, the head of Western Fuels publicly stated, apparently seriously, that he'd welcome CO_2 levels three and a half times their natural level, the thought of which triggers nightmares in most climatologists. And Western Fuels is not troubled by rising temperatures, either, pointing out that "warm is good, cold is bad." This particular wing of the coal industry is simply taking the notion of subduing nature to its logical extreme: the creation of a planetary greenhouse.

The major coal companies that increasingly dominate the industry have been unwilling to endorse openly such planetary reengineering, or the idea that increasing CO_2 is a net positive.** Indeed, they don't need to embrace such radical views. They have found substantial success by using more traditional political means.

---

*Western Fuels' faith that more CO_2 is automatically better for all plants, and therefore all ecosystems, is widely rejected by the scientific mainstream. For example, outside of greenhouses, other nutrients—such as nitrogen—are limited, so the CO_2 stimulus is much smaller. Moreover, because plants grown with more CO_2 have shown reduced levels of protein, more would need to be eaten for the same nutritional value. Also, some plants, including some fast-growing invasive species, respond more to CO_2 than others; this means that the balance of plants in a given ecosystem (and of the animals that depend on them) would be further destabilized, posing yet another challenge to biodiversity.

**Western Fuels established a nonprofit group called The Greening Earth Society to promote the idea, but it has been able to garner only anonymous supporters. The group's communications advisor stated that although he believes many members of the coal industry appreciate the work the society does in pro-

The U.S. coal industry had a lot at stake in the 2000 presidential election. In the words of one coal company executive,
"Albert Gore was trying to put us out of business." Global warming had become one of Gore's defining political issues ever since
his 1993 book *Earth in the Balance* raised awareness about it.
Gore was closely associated with the Kyoto Protocol, which he
helped negotiate. The highly motivated coal industry tripled the
level of its campaign contributions from the previous presidential contest, giving generously to George W. Bush, a former oil
man from Texas who was already predisposed to share their
environmental views.

For the most part, neither climate change nor coal were factors in the 2000 election, except in one place, where they were
critical: West Virginia. The Bush campaign did not believe it
had any chance of winning the strongly Democratic state, but
the West Virginia coal industry persuaded them otherwise. The
issue was not just climate change; the state's mountaintop-
removal mining practices were under legal challenge, and the
Clinton-Gore administration had come down mainly against the
industry. The *Wall Street Journal* has detailed how the state's
coal industry took the lead in raising unprecedented sums of
money and support for Bush, and ultimately delivered the state

---

moting the advantages of rising CO_2 levels, "they don't want to be associated with
us." However, Fred Palmer, the long-time head of Western Fuels and the source of
most of the quotes above, recently became head of legal and political affairs for the
world's largest coal company, Peabody Energy. Palmer says he still holds the same
views, but that Peabody does not necessarily endorse them.

by 52 percent. According to the *Journal*, top White House staffers agreed that "it was basically a coal-fired victory." If Bush had not won West Virginia's five electoral votes, Gore would have been in the White House.

During Bush's first five months in office, he issued a pro-coal energy plan, took much of the momentum out of the EPA's enforcement efforts against coal plants, eased up on mountain-top mining restrictions, and, most important, came out firmly against Kyoto or domestic CO_2 controls. The head of the West Virginia Coal Association called it "payback," noting that Bush had told him "that he is appreciative and that he knows if not for us, he wouldn't be president." In most other developed nations, coal use has plummeted dramatically in recent years, dropping 42 percent in the last fifteen years in western Europe and even more in Britain. In the United States, though, the industry is celebrating a resurgence. The coal industry trade press blares out rapturous front-page headlines like "King Coal Is Back!" and "Another Record Year for Coal." The United States burned more coal than ever in 2001, and there is even better news for the industry. For years, there were next to no new coal plants built in the United States; the growth in coal use came entirely from running existing plants harder. Then coal won the triple crown: Bush was elected, the price of natural gas briefly but dramatically spiked, and California's electricity crisis scared the nation. Today, new coal-fired power plants are in the works in many states, each with a potential operating life of many decades. By one government count, there are some ninety new coal plants now being planned in the United States, which if actually built will cost some $64 billion.

SINCE THE OIL SHOCKS of 1973 and 1979, coal has been justifiably celebrated as the most American of fuels—a reliable source of domestic energy that made the nation less vulnerable to the messy world outside its borders. When oil prices skyrocketed, Americans began to call their country the "Saudi Arabia of coal," and to call coal the nation's "ace in the hole." Some power plants that had been burning oil were required by law to convert to coal, and all new oil and gas plants had to be able to burn coal as well. Air quality laws were loosened or delayed, the federal government pushed coal as the centerpiece of its energy strategy, and in many circles coal became the emblem of American energy independence.

This not-so-distant history makes international efforts to limit greenhouse gases doubly galling to many coal advocates. Not only is coal use being threatened yet again by environmental concerns, but this time at least in part by foreigners with environmental concerns, turning the issue into not just a question of environmental protection but of national sovereignty. The recent head of the National Mining Association, which counts most of the coal industry among its membership, has called ratifying the Kyoto Protocol nothing less than "unilateral economic disarmament." Under the treaty, "for the first time in history, the United States would allow a foreign body, dominated by developing countries, to restrict and control the U.S. economy.

United Nations bureaucrats would decide where business will invest and where jobs will be developed and U.S. sovereignty will be surrendered."

In fact, the treaty has no such provisions, but Kyoto does genuinely threaten coal use, and the quote does highlight the uncomfortable but unavoidable issue underlying the climate change debate: the relationship between the rich and poor nations of the world. One of the reasons the Bush administration rejected the Kyoto Protocol, indeed one of the first arguments always made by the treaty's opponents, is that it simply isn't fair for the United States and other developed nations to reduce their greenhouse gas emissions when under the treaty developing nations are still allowed to increase their emissions without limit. The Senate, even before Kyoto was negotiated in 1997, formally resolved that it would not ratify a deal that did not include substantial involvement from third world countries. The coal industry and other critics of the Kyoto Protocol argue that American jobs would be lost to the third world. They also point to reports that greenhouse emissions from the emerging economies are climbing at a faster rate and could surpass those from the developed world in the decades ahead.

The developing world sees the fairness issue quite differently. The United States alone accounts for one quarter of the world's carbon dioxide emissions. The rich nations together account for more than two-thirds of carbon dioxide emissions, and yet the lion's share of the costs may well fall on the poor nations. Emissions in developing countries may be rising, but only because these nations are trying to do what their rich cousins have already done—build a modern economy using fossil fuels. On a per capita basis, their greenhouse gas emissions are still extremely modest. They therefore ask the rich nations to use the wealth they've already extracted from fossil fuels and take a decisive lead in solving this problem without halting the long-awaited industrialization of the poorer world.

When Americans list the developing nations exempted from Kyoto's restrictions, the nation that tends to lead the list is China, and for good reason. Not only is China the largest nation in the world but it will likely become the world's largest economy in the decades ahead. China burns more coal than any other nation. Although China consumes far less energy in total than the United States, it gets roughly two-thirds of that energy from coal, just as the United States did around 1925. This is not to suggest, though, that China came late to coal; as we shall see, it began dipping into its vast coal reserves centuries before western nations dipped into theirs. Coal has played a surprising role in China's long and astonishing history—particularly in its competition with the West. Today, China illustrates the lure and threat of coal more vividly than any other nation in the world.

eight

A Sort of Black Stone

IF YOU GET ON A TRAIN in Beijing and travel a couple of days west through the mountains and toward the center of the vast Asian landmass, you will find yourself in Inner Mongolia's forbidding Ordos Desert. I made this journey not long ago. As I looked out the train window I was struck by the emptiness of the land, a reminder of how far I was from the crowded bustle of coastal China. There was the occasional lone herdsman guiding a few forlorn-looking sheep. It was a mystery to me how the few tufts of grass and brush clinging to the yellow sand and rocky hills could yield enough energy to support any animals at all. Here and there the smokestacks of isolated factories rose up out of the desert, but it was no mystery where they got their energy. It was from the jungles that once grew here and that bequeathed to China coal deposits richer than those of any other nation except the United States.

I was traveling with a young English-speaking Chinese woman from Beijing whom I will call Jane. Not only did Jane prove to be a highly capable guide and translator but she had a charm that could get us into places I could never have otherwise seen—and when needed, get us out again.

One such place was a coal-burning power plant outside a small city in the Ordos. This plant used the same basic generating technology as the Sherco plant in Minnesota, and was about the same age, but it was far smaller, and already slightly bedraggled. A window had been broken and left unfixed. So many interior lights had burned out that it was hard to see the electric generator. I saw more people at this small plant than at the much larger American one, but none appeared to be working at the time. Rather, they gathered in groups of four and five to watch me, smiling and occasionally shouting out an English "hello." They laughed heartily when I tried to respond in Chinese. Jane explained that although two Western men had visited the city years before, I was the first Western woman most of these people had ever seen in person.

Remote as it is, this corner of Inner Mongolia, like the rest of China, is ready for the outside world to discover it. The city's top hotel may have hot water for only one hour a day (by prearrangement, so they can crank up the coal-fired water heater), but behind the front desk a bank of clocks helpfully gives the time in major foreign cities such as London and New York. The power plant official who showed us around had been told only that I was an American tourist with an unusual fascination for coal, which was true enough as far as it went. I wondered why he even let us in, but then it became awkwardly clear that he believed I represented foreign investors who might be willing to pump some much-needed capital into his facility.*

As we left this plant, our taxi driver said he had a friend at another power plant we could visit a few minutes away. When we arrived, our driver boldly led us through the gates and onto the grounds. This was not just a power plant but also a coal-washing facility. In one corner it featured a striking black mountain of coal residue surrounded by pools of standing black water, and in another a grey mountain of ash. Before we could find our driver's friend, we heard shouting. A man of evident authority ran up and began yelling at me in Chinese. As I looked around nervously, imagining our imminent arrest, Jane smiled and reassured him with some friendly banter that I was harmless, and he calmed down. We left quickly. Jane told me later that the man had accused me of being there to document pollution to "embarrass China" for political reasons. It would have been hard to explain to him the far more complicated truth.

The very different reactions to my presence at these two power plants reflect the conflicting feelings the Chinese have toward the outside world. Years of isolation and the economic disasters of decades past have brought a genuine eagerness to connect with other nations along with a strong reluctance to be judged by their standards, especially environmental standards that rich nations did not embrace until they were rich. These are

---

*Today, some 10 percent of the investment in China's energy infrastructure is foreign.

sentiments best understood in the larger context of China's ancient glories, its humiliation at the hands of the West, and its desperate twentieth-century attempts to modernize—aspects of China's history that prominently feature coal.

SOME 5,000 YEARS AGO, one of the world's first civilizations took root along the Yellow River when the early Chinese stopped their hunting and gathering, sunk their Neolithic tools into the soft soil, and began to farm. Over the millennia the Chinese pushed farming as far north as the rainfall allowed; past that line only barbarian nomads could survive. The northern boundary of civilization was thus determined by the climate; the Great Wall was just China's way of reinforcing the natural border between wet and dry, settled and unsettled.

Climate not only drew the borders of Chinese civilization but promoted the rise of one of its most salient features: powerful, centralized government. Even south of the Great Wall, rainfall is so variable that China has always been buffeted between the twin disasters of drought and flood, and many millions have been killed. In Chinese mythology, the first dynasty was founded around 2200 B.C. by a figure who offered some relief from these calamities. Yu the Great—part god, part water engineer—brought peace between the earth and the waters and made cultivation possible by building dams, canals, and irrigation ditches (and also, legend has it, by dancing a special flood-control dance).

For thousands of years, China's real-life emperors had the critical responsibility of building and maintaining massive water projects, and they mobilized armies of people in the effort. The stakes were high for the emperor: Failure to control floods and droughts raised the question of whether his heavenly mandate to rule had lapsed and his dynasty should end. Success kept the crops growing and the taxes pouring in. The emperors succeeded often enough that China thrived and developed in ways that would stun the West.

Indeed, when around the year 1300 the traveler Marco Polo wrote his famous book describing the marvels of China to his fellow Europeans, he was widely condemned as a liar. Medieval Europe could not believe his descriptions of China's huge cities, the magnificent canals, the tremendous wealth, and the complex system of governance. Among the small oddities he described was this:

> Throughout this province [of northern China] there is found a sort of black stone, which they dig out of the mountains, where it runs in veins. When lighted, it burns like charcoal, and retains fire much better than wood; insomuch that it may be preserved during the night, and in the morning be found still burning. These stones do not flame, excepting a little when first lighted, but during their ignition give out a considerable heat.
>
> It is true that there is no scarcity of wood in the country, but the multitude of inhabitants is so immense, and their stoves and baths, which they are continually heating, so numerous, that the quantity could not supply the demand. There is no person who does not frequent a warm bath at

least three times in a week, and during the winter daily, if it is in their power. Every man of rank or wealth has one in his house for his own use; and the stock of wood must soon prove inadequate to such consumption; whereas these stones may be had in abundance, and at a cheap rate.

By this time, coal was already being used industrially in Britain, but Polo, of Venice, was obviously unaware of it, and he had no name for the mysterious stones.

The history of coal use in Britain and China has many parallels, but China was many centuries ahead. Stone Age people in northeastern China were using coal 6,000 years ago. The early Chinese, like the Romans in Britain much later, valued coal not because it could help them survive the bitter winters but because they thought it was pretty. Apparently they passed over the duller and more combustible outcrops of coal to seek out the rare, carvable pieces of jet. One ancient Chinese site has yielded expertly crafted pointed objects that look amazingly like small black golf tees, that archaeologists believe to have been Neolithic "ear-piercing ornaments." By the third century B.C., jet carving was a flourishing cottage industry over large parts of China.

It was also probably around this time that the Chinese first began burning their coal. In so doing, they quietly became the first human society to meet its daily needs by using solar energy that had arrived on the earth in the distant past. This new energy source would play no small part in the development of the sophisticated economy and society that would so impress Marco Polo.

China took a tremendous early lead over the West in large-scale iron production, mainly relying on charcoal. A Chinese text from 120 B.C. already complains about deforestation caused by metallurgy, a problem the English wouldn't complain of until the seventeenth century. (Interestingly, this same ancient Chinese text also bemoans the decline of primitive simplicity and the rise of unnecessary industries). When the deforestation around the blast furnaces became acute, around the eleventh century A.D., China solved the problem the same way Britain would much later: It figured out how to bake the impurities out of coal to form coke. Chinese blast furnaces then moved from the deforested hilltops to the coal fields, and the remarkable expansion of the iron industry continued.

When Britain began using coal to make cheap iron in the 1700s, it spurred that nation's urbanization and industrialization. When the United States did the same a century later, the industrialization of the eastern states took off. When China began using coal to make cheap iron in the eleventh century, something very similar happened; even in this largely premechanized age, coal and iron spurred industrial development on a scale that the world had never before seen, and would not see again until Britain's industrial revolution. Eleventh-century China's blast furnaces were run by private industrialists and manned by hundreds of wage laborers; they were fueled by coke, and they churned out thousands of tons of iron yearly. Iron works in one small region yielded some 35,000 tons of pig iron in the eleventh century, more than the amount produced in Great Britain in the early 1700s. One massive operation employed nearly 3,000 workers, including at least seven hundred coal miners.

This unprecedented level of Chinese industrialization took place during China's famed Northern Song dynasty, from 960 to 1125, when Europe was in its dark ages. The imperial capital was Kaifeng, a phenomenal city largely forgotten by history. It was, in the words of one Western historian, "a multifunctional urban center quite possibly unsurpassed by any metropolis in the world before the nineteenth century." Perhaps over a million inhabitants called Kaifeng home, including at least 150,000 guard troops.

Holding the empire of the Northern Song together took a lot of coal-smelted iron. In 1084, the government supplied 35,000 swords, 8,000 iron shields, and 10,000 iron spears, all to defend just one part of the empire. At just two arsenals, 32,000 suits of armor in three standard sizes were produced each year. A bow-and-arrow works near the capital made over 16 million bows, arrows, and steel arrowheads every year. The empire employed a hundred artisans to make "horse-decapitating swords" to defend against the horseback-riding nomads constantly threatening to invade from the north.

Keeping the residents of Kaifeng supplied with fuel was no small challenge. Early in the eleventh century, they still burned charcoal, but it was getting expensive. In 1013, the emperor offered charcoal from government supplies at half price; people were reportedly trampled to death in the hysterical rush to buy the fuel, and troops had to restore order. Eventually, the citizens of Kaifeng did what the industries had already done—they began to burn coal. Early in the 1100s, it was said that coal was so prevalent that no dwelling burned wood anymore. Kaifeng had completed the process that a much smaller London would go through five hundred years later.

China continued to thrive technologically for centuries more, despite repeated invasions. Then, in the mid-1400s, just when western nations were getting ready to explore and exploit the wider world, China turned inward, restricting foreign trade and contacts. Focused on resisting overland threats from the north and west, China let its powerful navy wither. Merchants caught trading with foreigners were executed, and for a time it was illegal to study a foreign language or to teach Chinese to a foreigner. Historians groping to explain how China lost the lead it had held for so many centuries often point to this self-inflicted isolation as the reason. Whatever the cause of China's decline, when westerners returned to China in the 1600s, the balance of power would be very different than it was in Marco Polo's day.

SURELY NO NATION ON EARTH has as many coal miners or coal mines as China. In 1996, 5 million Chinese mined coal, virtually all of them underground. At the same time in the United States, about 90,000 miners were digging about the same amount of coal. The reason for the disparity, of course, is that Chinese mines rely much more on cheap labor than on costly machines. In addition to its many large mines, China has tens of thousands of tiny mines that each employ just a handful of miners.* The small mines are vastly more deadly than the big ones, which are themselves quite dangerous. In 1991, a particularly bad year, 10,000 Chinese coal miners died in accidents. By comparison, the number of Americans killed in coal mining accidents in 1992, a bad year for the U.S. industry, was fifty-one.

I had not yet learned these statistics when I went into one of China's small mines in the Ordos, but it was obviously not a safe place. On the day Jane and I visited, the constant, gritty wind made it hard for me to keep my eyes open, and after a few minutes I could feel the sand grating between my teeth. From a distance, about all I could see of the coal mine was a spindly platform jutting out at an angle from the empty yellow mountainside. At one end of the platform, built of thin, rough-hewn logs imported from some distant place where trees still grew, a red flag flapped loudly from a wooden pole. A pair of thin metal rails ran along the top of the platform; a coal car ran along these rails and then disappeared into the mouth of the mine.

We followed the shift manager through the opening. The tunnel was so low that we had to bend to enter it, and so steep and slippery with sand and coal dust that we needed a handhold as we climbed down. I was tempted to reach up and grab the irregularly placed timbers forming the tenuous ceiling, but, conscious of the weight of the mountain above, I gripped the

---

*In 1998, China had about 75,000 mines employing an average of thirteen miners each. These small mines have a death rate seven times higher than the large ones.

cold stone walls instead. As the tunnel narrowed even more, we had to be careful in the darkness not to step on the cable that raced along near our ankles and that occasionally jumped a few inches to either side as it lowered the coal car into the tunnel ahead. This car and the surface engine that powered it were the only signs of mechanization, and they posed a constant threat. If by some failure of communication the coal car approached while we were in the tunnel, there would be no way to stop it and no room to dodge it. Toward the bottom, our guide decided we had gone far enough, and we scrambled, at times on all fours, back up to the tiny square of light that was the entrance. I was content to interview the miners on the surface instead of underground.

Later that day when the miners emerged from their shift, we were waiting. Perplexed by my presence and my interest, some of the exhausted miners were nonetheless willing to describe their work to me. I learned that about fifteen miners per shift walk into the mine down the treacherous slope at its mouth. From the bottom, it takes a miner about twenty minutes, walking in a crouch, to reach the portion of the coal face where he will work. There he carefully places his explosive, sets the fuse, moves to a spot of relative safety, and triggers a blast that will rip loose the coal. Then he makes his way through the dust and shovels the loosened coal into the car to be hauled to the surface. The miner stays below for his full eight-hour shift, blasting and shoveling by the weak illumination of his flashlight. He does not eat, and he does not stand upright, until his shift is over. One miner who has worked in this mine for many years explained to me, with a laugh, that in time you get over your fear and you get used to the ache in your back.

It is risky to assume that this is a typical Chinese coal mine. Some Chinese mines employ thousands of miners and are relatively modern. Even in centuries past, mining conditions varied enormously from mine to mine. A source from the late 1600s or early 1700s reported that "some galleries are twice the height of a man, some just high enough for a man to pass, some just wide enough to squeeze through sideways." Many mines were accessed by a sloping tunnel like the one I entered, but others by way of a vertical shaft. In the latter, miners typically rode into the mines on a stick tied to a rope, which was connected to a windlass at the top, often operated by four or more men. Light was provided by oil lanterns, and according to at least one source, the lanterns were sometimes fastened to the miner's head by his pigtail.

In the early twentieth century, as eight-hour shifts were introduced to the modern Chinese mines, shifts at other mines were twelve, eighteen, twenty-four hours long, or even more. In one remote province, the difficulty of raising and lowering men with the equipment that also raised water and coal led to what they called "the big shift," which required the men to live in the mines for as long as fifty-five days. This might explain the oddly domestic underground scene that a Western visitor in the early twentieth century reported: "A particularly weird site was a big room, 60 ft. in diameter by 30 ft. high, near the shaft, where a monstrous fire was glowing; on a line stretched between two poles some rags of clothes were drying, and men were lying all around, some asleep and some smoking opium. Another curious feature was the large number of rats, fed with millet by the miners, and believed to give warning of accidents." Miners in Pennsylvania were at this time forming the same bond with mine rats, and for the same reason.

Back in the 1600s, when the Scottish mines were using slave labor, some Chinese mines were too; but in China, the practice was illegal. Even so, in rare cases slave labor in the mines continued into the 1900s. Once, a government official on an incognito inspection tour of a remote interior province was kidnapped by a mine owner and kept underground as a slave. The official "bit his finger to draw blood and every day wrote his name and rank on a large piece of coal, hoping it would be seen and he would be rescued." It was three years before troops were sent to free him. Children also worked in some Chinese mines, but probably fewer than in early English mines. Women generally did not mine coal, not because of gender privilege but because footbinding, common in northern China into the early twentieth century, left many of them partially crippled and unable to perform heavy labor.

Although the traditional Chinese mines varied widely in size and conditions, they were all limited by the lack of steam engines to pump water from the mines. The Chinese typically fought the seepage of water into the mines with leather buckets and human muscle, then abandoned the mine and opened another. Even with lots of cheap labor, the absence of steam engines and railroads meant that Chinese coal was more expensive in some coastal Chinese cities than coal imported from Japan, Australia, or even Britain.

The buyers of this imported coal were largely British traders who had pushed their way by force into China in 1842 after the first Opium War. Later, other Western and Japanese merchants arrived to take what they could from the weakened nation. The foreigners who had once so admired China's technological prowess now used their own to force the Chinese to give up their sovereignty over various Treaty Ports, like Shanghai, in which foreigners established commercial and industrial enterprises hungry for coal. The foreigners were eager to tap into China's rich coal veins with their modern machinery, and the worried Chinese were eager to stop them.

When the Chinese decided that they would need to emulate western technology if they were to defend against it, they launched what was known as the "self-strengthening movement" of the 1860s and beyond. This defensive drive to modernize was spearheaded largely by Viceroy Li Hongzhang, one of the most influential Chinese officials of the late 1800s. He argued that Britain, a small nation, had prospered in the production of coal and iron and was now the richest and strongest nation on the earth, pointing the way for China. Despite China's long dependence on coal, though, mines were often opposed on grounds that they would harm an area's *feng-shui* (the wind and water spirits). Li's mine complex, named the Kaiping mines, had to overcome fears that it would anger the earth dragon and disturb imperial tombs sixty miles away. Eventually, though, the coal shafts were sunk. Apart from Li's arsenals and shipyards, his Kaiping coal mine complex would become "the first successful, large-scale effort to introduce Western technology and methods into industrial production in China."

But transportation remained a problem. Over the centuries, coal in China had been moved largely by pack animals. Beijing was kept warm by long parades of "sooty and weary camels" bringing coal from the west. Kaiping needed rail transport, but the imperial government hated railroads; instead, a seven-mile-long mule-driven tramway was proposed. Then a British engineer hired by the mine was allowed to build, secretly, a locomotive out of a steam engine bought to lift coal. Known as "the Chinese Rocket," after George Stephenson's Rocket, the locomotive was revealed in June 1882 when it pulled a party of officials down the track at twenty miles per hour. According to the Chinese Rocket's builder, "Nothing more was said about mules and the stables built for their use were quickly demolished." Somehow Li had forced the railway's acceptance. Kaiping not only introduced modern mining into China but also brought about the long-delayed birth of China's rail system.

As the century came to a close, Kaiping was running short of capital and turned to a British mining promoter, C. Algernon Moreing, for help. Moreing also helped place one of his talented young employees from the United States into the provincial Bureau of Mines to advise China on mining policy and, incidentally, keep a keen eye on Moreing's investment at Kaiping. The man was Herbert C. Hoover, who was one of the world's most renowned mining engineers long before becoming one of America's most reviled presidents. Hoover arrived in

China with his new wife in 1899, just in time to settle in before the Boxer Rebellion.

The Boxers were desperate peasants who called themselves the Righteous Harmonious Fists, and they combined a potent mix of martial arts, trance-inducing rituals that they believed made them bullet-proof, and a passionate hatred of foreigners. Since foreign nations were eagerly carving China up into spheres of influence, there was plenty to be angry about— though the killing frenzies of the Boxers in 1900 were largely fueled by sensational rumors of, say, missionaries eating babies and other imagined evils. An American missionary in Shanxi Province wrote in his diary that many Chinese believed that westerners were controlling the weather by fanning away clouds and perpetuating a famine-inducing drought. He and his wife would be among the many missionaries murdered that bloody summer.

The wave of violence against foreigners prompted the troops of several nations to race to China and occupy significant portions of it. Parts of the Kaiping mines had been seized by the Russians. Hoover managed to convince Kaiping's Chinese management that to keep the mine out of Russian hands, it should be put under the British flag. The company's assets were temporarily conveyed to Hoover himself, as Moreing's agent. The history-making Kaiping mines—proudly launched to strengthen China against foreign aggressors and now the largest industrial enterprise in the country—were dropped into the lap of a twenty-six-year-old American engineer.

Hoover soon became general manager of Kaiping, and the Chinese managers were crowded out. A British diplomat stated, undiplomatically, that the Chinese had been "fleeced" and "got themselves fairly had by a Yankee man of straw acting for More-ing." When Chinese officials failed to get the mines back through the British courts, they were furious. The loss of Kaiping eventually became a cause célèbre among Chinese nationalists.

Britain continued to dominate these and other Chinese mines in the decades ahead. Meanwhile, the coal-poor Japanese developed the rich coal regions of the northeastern part of China that the West calls Manchuria. Manchurian coal gave Japan an additional incentive to invade China in the 1930s; it also helped fuel that invasion and the rise of Japanese industrial and military power more broadly, setting the stage for the global conflict ahead.

China would be at war with Japan until 1945, and then at war with itself until 1949. When the Communists came to power, they were determined to rid the nation of foreign power and privilege, including that at the coal mines. Mao Zedong would have shared the view of the U.S. coal industry that a threat to national coal supplies was a threat to national sovereignty, and with a much stronger historical basis for that view.

JUST SOUTH OF THE GREAT WALL in China's Shanxi Province, there is a magnificent shrine with thousands of sculptures carved into a sandstone cliff. Most of the figures are Buddhas, and some are giants, over sixty feet high. Over fifteen centuries ago, when the shrine was carved, this area was briefly the center of Eastern culture and civilization; today, it is the center of China's coal industry. These ancient Buddhas sit smiling mischievously on top of one of the world's richest deposits of coal. When Jane and I visited the shrine one cold March day, a large heap of mine waste across the valley had caught fire and the plume of smoke was slowly spreading across the sky. The Buddhas sat gazing directly at the smoke, a thin dark layer of corrosive coal dust coating their laps and shoulders.

In some ways, Shanxi is the Pennsylvania of China: lots of anthracite in the east, lots of bituminous in the west, and mountains between the coal and its main markets on the populated coast. In Shanxi, though, nature has added another barrier: deep layers of loose, windblown soils known as "loess." Loess soil is found all over the world, but nowhere as much as in northern China, where it can be as deep as 250 feet and forms the Loess Plateau. Loess also gives the region its bizarre otherworldly landscape by eroding in a strange vertical fashion, leaving behind deep ravines with high steep walls. These ravines made it easier to find the coal, but they undermined attempts at building roads and paths, making it nearly impossible to transport the coal. Until relatively recently, cargo in some parts of the province had to be carried on the backs of mules, camels, and people. The loess kept China's greatest stash of coal largely unexploited before modern times.

Today, the twisting and potholed roads in the hills circling the shrine and the nearby city of Datong bustle with activity.

Innumerable little blue coal trucks, filled to the brim with chunks of coal sometimes as big as one's head, veer around heavily loaded donkey carts, people on bicycles, and the occasional pig rooting through the roadside litter. Along one of these roads a colossal statue of Chairman Mao looms over the offices of a coal company, his arm raised as if in benediction of the coal fields. There is no hint of the tremendous disruption Mao's policies caused to the coal industry, to the coal regions, and indeed to all of China.

In the summer of 1958, Mao launched his Great Leap Forward, plunging the nation into what has been called "the biggest and most ambitious experiment in human mobilization in history." Nikita Krushchev had just announced that the Soviet Union would surpass the United States in major commodity production, including coal, in fifteen years. Mao countered that China, tapping into its vast rural population, would surpass Great Britain's production levels in even less time. China, Mao decided, would finally catch up to the West and develop its economy with a tremendous explosion of manpower and will power.

Some 500 million peasants were organized into large rural communes. Tens of millions were put to work manually building dams and reservoirs, often without the guidance of engineers, who were condemned as bourgeois experts. And, in a move that turned specialization—a hallmark of the industrial revolution—on its head, the communes were forced to become self-sufficient in as many kinds of production as possible. They set up millions of small industrial enterprises, including thousands of small coal mines and the operations that came to symbolize these strange times: the "backyard steel furnaces."

Mao wanted "the whole people making steel," though he understood little about what went into the process. Within weeks, primitive brick furnaces rose up in agricultural communes, at factories, at hospitals, and in the playgrounds of many of the nation's schools. Peasants, factory workers, doctors, and schoolchildren tried to help meet the party's steel production targets.* As many as a hundred million Chinese—roughly twice the entire population of Great Britain, the nation whose steel production they were trying to top—were feverishly employed in the feeding and tending of an estimated 1 to 2 million little furnaces, some of them built in a matter of hours.

In putting the masses to work in this way, Mao had overlooked many things, including the huge amounts of coal that steel production requires. Valiant efforts were made to expand coal production to help meet demand. Small new coal pits suddenly appeared in the countryside. By the end of 1958, by one estimate, some 100,000 coal pits were in operation, worked by some 20 million peasants. Coal was also mined at relatively large mines, and not always by "free" labor. It has been estimated that millions of political prisoners were held in labor camps during this time, and some of them were mining coal; according to one account, 100,000 prisoners, half of them Tibetan, were sent to mine coal in one remote western province.

---

*Actually, they were mostly trying to produce pig iron, which would theoretically be turned into steel at steel works.

As the steel campaign continued, the coal problem became clearer. Mao was advised that 600,000 men would be needed just to clean the coal needed for steel production. And, moving the coal and iron ore to the million scattered furnaces created a transportation nightmare. Mao later admitted that he had failed to predict how the nationwide steel campaign would suddenly overload the nation's rail system with coal and ore, much of which ultimately had to be moved by hand-pulled carts. Chinese dissident Ni Yuxian was growing up in a village near Shanghai during the steel drive; his biographer reports that he was assigned by his middle school to a brigade of children who gathered coke from a steelyard ten miles away and hauled it to their playground furnace. It was an eleven-hour round trip, with two children pulling each heavily-loaded cart.

At least these furnaces had access to coal. Although this was a period of soaring coal production, it was also a period of coal shortages as the coal was funneled into the steel drive (and into increased electricity production). Because factories and homes had to burn wood, as did the citizen-steelworkers when they couldn't get coal, the surrounding countryside was sometimes stripped of its already sparse tree cover. There are even reports of wooden doors being removed from homes to help fuel the steel production.

While Mao and other party leaders entertained the fantasy that this "steel" would make China strong, many of those sweating over the molten metal knew that the lumpy globs created in their furnaces were useless. Jung Chang, who, as a six-year-old girl in Sichuan, helped gather scrap metal for her school's iron campaign, has written that the people called these lumps "cattle droppings." China, whose advanced iron production methods in ancient times had put it centuries ahead of the West, was now dissipating the energy of its people—and its iron and coal—in the production of metal unfit for use. Later generations would inherit heaps of congealed metal in their playgrounds and court-yards where furnaces had been abandoned.

The Chinese paid a horrific price for this massive but ill-planned paroxysm of work: One of the greatest famines in human history killed tens of millions of people from 1959 to 1961. The peasants bore the brunt of the famine; some were reduced to eating leaves, bark, grass, and scraps of cotton. The economy was in a shambles and would take years to recover. Mao and the party leadership declared that the Great Leap Forward was over and moved on to other, more modest campaigns to address the nation's severe food and fuel shortages.

As for coal, after soaring briefly during the Great Leap For-ward, production crashed again. In the late 1960s, Mao decided that, rather than burden the trains by shipping coal from the north to the more densely populated south and east, China should just dig for coal where the people lived. The meager coal seams of the populated areas made this a deeply irrational policy, but Mao's decision was surely not subjected to much debate. These were the years of China's calamitous Cultural Revolution. The Red Guard, largely made up of millions of Mao-intoxicated teenagers waving the little red book containing the chairman's quotations, were dragging intellectuals and disfavored officials through the streets. Challenging Mao had never been more dangerous, and so much of the northern coal industry was uprooted, and mine exploration and construction companies were moved to the more populated regions.

The skimpy reserves in those areas could not yield enough coal to support the kind of large, modern mines the nation needed, so the industry rebounded slowly. When Mao died in 1976, the Ministry of Coal began to undo his policy, shifting the industry back to the coal regions of the north. Hundreds of thousands of workers and their families had to be relocated, but finally they were digging in the right place.

The turmoil is not over for the Chinese coal miner. As they are in the United States and other coal-producing nations, the small inefficient mines are shutting down in favor of larger ones. In China, though, the scale of the disruption is mind-boggling: Beijing claims to have closed down 30,000 small mines just since 1998. Although the true number is surely less, there are undeniably painful reforms underway that have already thrown perhaps a million Chinese coal miners out of work.* These sweeping changes reflect the fundamental shift in Beijing's economic philosophy over the years: In a move more reminiscent of J. P. Morgan than Mao Zedong, the Communist government is now openly urging

---

*According to widespread reports, many communities have defied Beijing and quietly reopened their small mines; as a result, several officially "closed" mines have suffered deadly mining accidents in recent years. However, reports that miners are being laid off in huge numbers, including at large state-run mines, are more credible. Between 1992 and 1995, reportedly 883,000 coal miners (more than ten times the total U.S. coal mining workforce) were laid off, and there are plans to lay off nearly 800,000 more.

coal companies to merge into larger and larger enterprises, and to form "cartels" to limit overproduction and improve profitability.

CHINA'S RURAL POPULATION has prospered measurably since 1976. The two decades before Mao's death are sometimes seen as lost decades, but the two that followed saw China become not just the fastest growing economy in the world but one of the fastest growing ever. Per capita income virtually quadrupled between 1977 and 1997 under Deng Xiaoping's economic reforms, which let peasants grow and sell what they can in an open market.

Today, the lives of many peasants are a startling mix of old and new. In the hills of northern Shanxi's coal region, Jane and I visited a charming elderly couple who live in a cave. Since Neolithic times, people have dug homes out of the region's soft loess cliffs. The traditional cave homes (we would call them earth-sheltered) are increasingly rare, but thousands of Chinese still live in them. Today, like the home I visited, they usually have a front wall made of brick and glass, and arched interior walls to hold back the loess. The cave homes are quite comfortable, though utterly lethal when earthquakes strike.

The couple graciously offered me a seat on their *kang,* a traditional brick platform built around a coal stove and heated by pipes. The coal-fired *kang* has been at the heart of domestic life in northern Chinese homes for centuries. Carpets or mats are placed on top so that people can sit there during the day, and traditionally the family sleeps there at night. A Jesuit visiting Beijing in 1600 commented on the *kang*'s critical role in helping even the poorest families survive the bitter northern winters. While the British were still centuries away from accepting enclosed stoves, the Chinese were practically living on top of theirs. (This practice was not without risk, of course; the visitor in 1600 noted that "sometimes it happens that the Stove takes Fire, and that all that are asleep upon it are burnt to Death.") The *kang* I sat upon was warm and pleasant and not much changed over the centuries, except that above the *kang* dangled an empty electric light socket.

We also visited with this couple's more affluent neighbor across the village. When we arrived, she was frying vegetables in a wok on top of a small, coal-fired stove. References to the use of coal for cooking in China go back to the early seventh century. (A Chinese scholar wrote in 1637 that coal is an unsurpassed fuel for cooking all dishes except bean curds, which become bitter if cooked over a coal fire.) A skittish white sheep running in and out of this woman's tiny kitchen gave her home a decidedly rustic feel. At the same time, though, a large color television, suspended from the ceiling in the corner of the kitchen, was flashing pictures of the wider world. This woman, old enough to remember the famine of the early 1960s, could now plainly see the affluent images of rich nations commonly broadcast on Chinese television.

After Mao's death, Chinese dissidents were finally able to state openly what they dared not say before: China was suffering a long-term energy crisis that made the energy problems of developed nations pale by comparison. Peasants—who represent the majority of the population—had 20 percent less energy than minimum daily requirements. In some areas there was too little fuel to cook even one meal a day.

The government slowly introduced market forces into the energy arena, opened the door to foreign technology, and devoted itself to increasing China's energy efficiency. Coal production rose, and a tremendous boom in the construction of electric plants continued until very recently. Between 1988 and 1999, China increased its generating capacity over two and one-half times, the equivalent of adding one medium-sized power plant every two weeks. Like developed nations, China is trying to introduce competition into its electric-power markets, but it is proceeding cautiously. In 2001, it announced that it would slow down some of its reforms to avoid repeating the deregulation mistakes of California!

In 1999, China actually had a glut of electricity; the government put a three-year moratorium on the construction of new, coal-fired power plants while it built more wires to deliver the power. The glut is not likely to last long, though, considering how fast the economy is growing and how many Chinese still lack adequate supplies of electricity. Given the nation's recent past, it is amazing that almost 40 percent of rural Chinese households already have a television and over 10 percent have a refrigerator. And yet there is clearly plenty of room for growth in electricity demand. Per capita, the Chinese still use thirteen times less electricity than Americans. By 2015, China's generating capacity is expected to double over 2000 levels, and coal is on track to provide the vast majority of that new electricity. This is true even when factoring in the power provided by a modest

nuclear power program and by the hugely controversial Three
Gorges Dam on the Yangtze River, construction of which is taking China's long history of enormous water projects to a dramatic new level.

THE THREE GORGES DAM is designed not just to provide
energy but to control flooding—a problem that continues to
plague China. As recently as 1998, floods along the Yangtze
destroyed 5 million homes, killed over 3,000 people, and affected
a population nearly as large as that of the entire United States.
And the risk of flooding is just one of the reasons why China—
its dense population already stretching the land to the breaking
point—is particularly vulnerable to future climate changes.
Another is the menacing growth of its deserts. Much of the over-
grazed Inner Mongolian steppe is now bare sand, the source of
fierce dust storms that increasingly cover Beijing with a layer of
yellow-brown grit; or, if the dust falls with rain, with a slippery
coating of mud.* The Gobi is moving toward Beijing so rapidly
that, some predict, sand dunes could sweep into Tiananmen
Square within the century.

These dust storms threaten not only the Chinese but everyone downwind, including those who think they are protected by
an ocean. Researchers recently watched by satellite as two large
dust storms arose, picked up industrial pollution from China,
blew across the Pacific, and reached North America. The first

---

*In the 1950s, five strong dust storms hit the capital; in the 1990s, there were
twenty-three, and in the year 2000 alone, there were thirteen.

dust storm brought a visible haze to areas from Canada to Arizona. People hiking in the mountains at Aspen, Colorado, inhaled Mongolian dust and Chinese pollution still strong enough to violate U.S. air standards, and the cloud remained detectable all the way to the Atlantic. We've recently learned that, even between dust storms, Chinese pollution often drifts to North America. The greatest source of that pollution, of course, is coal.

From any one of the many new office towers in Beijing, you can clearly see the downside of China's heavy coal dependence—a grey haze of pollution hanging constantly over the city. Beijing's air pollution is far worse than that of most Western cities today, though not nearly as bad as the pollution in London, Manchester, or Pittsburgh seems to have been at its nineteenth-century worst. Even so, after I bicycled across Beijing, as millions of Chinese do daily, my lungs were stinging for hours.

It is now well known that, thanks to China's coal use, the Chinese people have been breathing some of the dirtiest air in the world. In 1995, five of the world's ten most polluted cities were in China, and Beijing was one of them. Some of the dirtiest smokestack industries, even steel mills, still release their plumes over crowded urban neighborhoods (incidentally, China has now achieved Mao's dream of producing more steel than any other nation). The problem is not just industrial, though. Hundreds of millions of Chinese still heat their homes and cook with coal; in the ancient brick neighborhoods of Beijing, honeycombed coal briquettes are delivered by bicycle through the alleys and burned in tiny stoves, and sometimes in barrels on the street over which woks have been placed to cook the family meal.

Recent studies by both Chinese and Western researchers suggest that China's air pollution causes perhaps a million deaths a year (compared to estimates of tens of thousands of deaths from air pollution in the United States). That figure would amount to one death in eight in China, and includes cancer and other deaths from the coal smoke that builds up inside homes from stoves. Millions more Chinese are inadvertently poisoning themselves by drying corn and chili peppers indoors over coal high in arsenic and fluorine, and they now suffer conditions that include skin cancer and severely deformed bones.*

Many of China's pollution problems are caused by SO_2, which the nation emits more of than anyone else in the world. Indeed, SO_2 not only contributes heavily to the death toll but gives southern China a huge acid rain problem, affecting some 40 percent of the land area. Even though more coal is burned in the north, the alkaline soils of the expanding deserts neutralize that acid as it falls; thus, northern coal emissions tend not to cause acid rain, at least not until they blow to Japan.

Now for the better news, less well recognized outside China. In the last few years, the Chinese government has been relatively frank about acknowledging its horrendous pollution problems, and even publishes its air quality monitoring data in

---

*More than 10 million residents of southwestern China suffer from fluorine poisoning, and thousands suffer from severe arsenic poisoning, obtained in this way.

the newspapers. Beijing has frequently declared its commitment to clean things up, and has followed up with sweeping, Chinese-style national campaigns. Highly polluting enterprises have reportedly been closed down by the tens of thousands since the mid-1990s. Although many of these facilities will probably reopen, as have some of the supposedly "closed" coal mines, this action still represents a significant shift in policy with great potential to improve China's air.

In some places the air in China is not just getting better, it appears to be getting a lot better. In Beijing, where officials have been investing hugely to improve air quality before the 2008 Olympics, sulfur dioxide levels were reduced by an impressive 41 percent just between 1998 and 2000 (for perspective, this means that SO_2 levels are no longer triple world health guidelines, but still nearly double them). Central to this effort is the move to replace urban coal fires with natural gas and electricity. And these initiatives are underway not just in high-profile cities like Beijing but also in places far off the beaten track. In one Inner Mongolian city I visited, anthracite imported from Shanxi Province was starting to replace the cheaper, dirtier local coal for urban use. Nationwide, though, pollution controls for sulfur dioxide at power plants are still quite rare, and SO_2 emissions are still rising. In short, China is just beginning to deal with its acute urban pollution problems, but is still a long way from addressing the chronic regional ones.

China's air pollution would be much worse if not for the work of people such as Professor Zhou Dadi, whom I visited in his Beijing office. He is not an environmental regulator, but rather the head of the nonprofit Beijing Energy Efficiency Center (BECon), and even the existence of such a center is both a surprise and a cause for hope. It was created in 1993 through a collaboration between the Chinese government, the U.S. Department of Energy, and the World Wildlife Fund, groups with very different agendas but a common interest in reducing China's waste of fossil fuels. Today, although BECon is mainly supported by the Chinese government, its all-Chinese staff works closely within a strong international network and has gained wide respect. When President Clinton visited China in 1998, Professor Zhou—whose career was once nearly destroyed by Mao's antiintellectual Cultural Revolution—was one of a handful of nongovernmental organization leaders he met with. On the day I visited, Professor Zhou's office was filled with visitors from around the world.

Like all developing economies, and like all centrally planned ones, China's economy is riddled with energy inefficiencies. BECon's work involves trying to reduce them by, among other things, encouraging government policies that increase efficiency, and pushing for the use of more modern technologies. As part of the larger Chinese effort to modernize energy markets and use, China has moved further toward energy efficiency than observers would ever have predicted in years past. In fact, for two decades, the nation's energy use has grown only half as fast as its economy, and at a much slower rate than energy use in other countries at similar stages of development. Not only has this kept additional pollutants like sulfur dioxide out of the air, but it has made China's contribution to global warming about half what it would be otherwise.

Western analysts were recently shocked to see reports claiming that although China's economy grew 36 percent between 1996 and 1999, its energy output fell by 17 percent, and its greenhouse gases dropped 14 percent. These dramatic drops were due almost entirely to plummeting coal use, amounting to a reported 30 percent decline over the last few years. Unfortunately, these numbers were based on notoriously flawed official statistics. Chinese estimates of economic growth are generally assumed to be exaggerated. Moreover, Beijing's recent efforts to force the closure of small coal mines and to reduce the use of high-sulfur coal have given local officials new incentives to underreport their coal use, and we know that many small mines continue to operate illegally. Even after factoring these things out, though, it is very likely that coal consumption fell significantly (perhaps by 10 percent after 1996) during a period of strong economic growth, and that carbon dioxide emissions fell significantly as well. It is certain that China is emitting far less carbon dioxide than earlier studies projected it would at this time.

What is the significance of this decline in Chinese coal use? It does not mean that China has turned a permanent corner. Coal use after 2000 appears to have picked up. Efficiency improvements will no doubt continue, but Chinese planners have made it clear that they still intend to expand their power sector enormously and to fuel that expansion mainly by using the centuries-worth of coal beneath their soils. Because per capita emissions are still so tiny, China claims the right to increase its greenhouse gases as it develops. As Zhou Dadi explained to me, "when we are richer, we will take more responsibility" for slowing climate change. In the meantime, though, China looks to the rich nations to take the lead in reducing carbon dioxide. Markets can promote global improvements in energy efficiency, but, in Zhou's words, "markets surely won't drive this decarbonization, so there needs to be some mechanism" to make it happen, and right now the Kyoto Protocol is the only one out there.

As China continues to grow, its coal use and carbon dioxide emissions are expected to track upward again. Still, China's surprising energy efficiency gains—and the considerable room left for efficiency improvement—do offer a little hope and a little time we didn't expect to have. If the world uses that time to aggressively develop alternative energy technologies, and quite possibly international financing structures, it might still be possible for China to change course before it locks itself into decades more of increased coal use, persistent air pollution, and rising greenhouse gases. The benefits, to China and to the rest of the world, would be tremendous.

nine

A Burning Legacy

WE'VE MADE A LOT OF MISTAKES over the centuries as we've struggled to understand the nature and impact of coal and its smoke. Some thought coal grew underground from seeds or in mines guarded by demons or dragons. Some saw in the mines scientific proof of the biblical flood. Some credited coal with protecting people from the bubonic plague; others accused it of promoting baldness, tooth decay, sordid murders, caustic speech, and fuzzy thinking. More recently, many of us believed we could burn vast amounts of coal indefinitely without disrupting the natural balance of the planet. No doubt we still have much to learn about coal, but at least we've been able to dispel many of the old myths.

There is, though, at least one truth that was more widely understood in the past than it is today—the critical importance of coal in shaping the fate of nations, and of the world as a whole. Coal transport lured the British to the sea, promoting the nation's growth from a small rural nation into a world-class commercial power. The Royal Navy was kept strong largely to protect the coal convoys; and in war time, it seized the coal ships and crews to fight its battles, helping Britain rule the seas. Thanks to coal, London grew into a metropolis large enough to become a vital center of commerce and cultural achievement. With an economic, military, and cultural influence far out of proportion to its size, this tiny nation began building a global empire of unprecedented reach, defeating native populations and European rivals such as France and Spain—nations with far more land and people, but far less coal.

And then there was the industrial revolution—fueled by coal, built around coal-smelted iron, and driven by two key innovations first developed to meet the needs of the coal industry: the steam engine and the railway. Coal alone did not make the industrial revolution happen any more than coal alone made Britain a global superpower, but neither event could have happened without it.

To grasp the magnitude of coal's global impact, we must try to picture history without the momentous, high-intensity pulse of industrialization that started in Britain and then swept the world. The mainly agrarian world would have stayed in place for decades or centuries longer, with slower technological progress, less material wealth, and more gradual social change. Mass-production capitalism would not have soared to prominence, industrial working classes and places like nineteenth-century Manchester would not have mushroomed, and the *Communist Manifesto* would never have been written. The North might have lost the American Civil War, or it might never have started, and the transformation of the American West would have happened slowly by wagon rather than quickly by rail. The World Wars might never have exploded without the industrial rise of coal-rich Germany. Colonial conquests would have been far less sweeping, dramatically altering the history of all the societies that were dominated by foreign industrial powers, including China's (whose ancient history would have been altered as well). The labor and environmental movements, if they had existed at all, would have taken very different forms. In short, none of the defining and epic struggles of the nineteenth and twentieth centuries would have played out as they did.

This is not to suggest the world would have been necessarily stable and peaceful, as a glance at our planet's violent preindustrial history shows. If human progress had been more dependent on harnessing surface energy rather than mineral energy, it's possible, for example, that slavery might have become an even more entrenched evil. And, although our air would have been cleaner and our climate less threatened, our forests and wilderness areas might have been more widely depleted. The pressure on the land would have been far greater because it would have been drawn upon for fuel as well as for food. No doubt, eventually somebody would have figured out how to turn heat into mechanical motion, inventing the steam engine or something like it, and the pressure on the remaining forests would have intensified. In such a world, heavily wooded nations like Sweden might have achieved global prominence.

Oil and natural gas resources would have been tapped, too, but probably much later than they actually were. Coal gave us the technological and industrial base we needed to exploit widely these harder to find, harder to move, and harder to control fuels. It provided, for example, the cheap iron and steel we needed for drills, pumps, tankers, railways, and pipelines. In a world following a flatter trajectory of technological progress, we might only now be starting to exploit petroleum by burning it in our kerosene lamps.

There are, in short, at least two very different paths civilization might have followed without coal. Humanity's technological and economic progress might have been so gradual that progress could have been more humane, allowing us to avoid much of the misery of the industrial revolution, and possibly even to develop environmentally sustainable ways of living. Or, maybe the greater pressure on the limited resources of the land would have simply led to a different series of wars and injustices, along with lingering poverty and a more complete consumption of the wilderness. Guessing at which path we would have taken without coal can at this point be little more than a parlor game. It may help us decide whether we think coal's influence so far has been a net blessing to humanity or a net curse; but ultimately that judgment will probably turn more on how we already feel about the glories and tragedies of our history and of our existing, coal-shaped world.

PEOPLE WILL PROBABLY always find some use for coal. Any substance versatile enough to pierce ears in Neolithic China, accessorize togas in ancient Rome, smoke out snakes in Dark Ages Britain, darken paint in prehistoric Pennsylvania, and transform itself chemically into goods ranging from pesticides to perfume, from laughing gas to TNT, probably has still undreamed-of future uses. But will coal still be a significant source of energy in the decades ahead?

Today, at least in the United States, coal advocates are banking on a new idea that would let us burn coal even in an age of climate change: carbon sequestration.* It involves capturing the carbon dioxide shooting out of coal plant smokestacks and permanently disposing of it. Although this is essentially the approach we've taken with most other air pollutants, the practical difficulties are so enormous that it has only recently been suggested as an option for CO_2. The U.S. Department of Energy, which is working on the concept along with the American coal industry, says that until the late 1990s, this technology was not even in the scientific lexicon.

Carbon sequestration would not be easy. Sulfur, for example, is only a tiny fraction of coal content, but we've spent decades and billions to catch the fruits of its combustion (sulfur dioxide), and we still haven't fully succeeded, even in developed countries. Because carbon is the *essence* of those trainloads of coal that

---

*The term *carbon sequestration* is confusingly used to refer to two very different processes. The first relies on growing more trees and other plants to draw the carbon from the air and store it in vegetation and soils. Although this form of biological carbon sequestration can help, it has serious limits: We're putting CO_2 into the air far faster than plants can absorb it, and these days most plants don't turn into coal but eventually just decay, putting their carbon back into the air. The second kind of carbon sequestration, relying on technology rather than plants, is the kind under discussion here.

power plants burn every day, capturing carbon dioxide presents technical challenges of a higher order of magnitude. Once captured, the massive quantities of CO_2 have to be moved to a disposal site, and, finally, disposed of in a way that won't come back to haunt us. It is, in a sense, like taking the enormous infrastructure currently devoted to the extracting, moving, and burning of coal and building it all over again, but in reverse. It can probably be done, but that doesn't necessarily make it worthwhile.

One problem is finding a permanent home for the CO_2. It is already pumped into oil wells where it helps flush out the oil, and it could also potentially be stored in abandoned coal mines, salt formations, and other geological structures. But it appears that not enough of those are available for the spectacular quantities of CO_2 involved, so researchers are eyeing the deep oceans. Unfortunately, though, when CO_2 dissolves it creates a mild acid that deep ocean creatures apparently find unpleasant.

There's also reason to wonder whether the CO_2 could really stay securely trapped at the ocean bottom. The potential instability of carbon dioxide dissolved in water was illustrated by the bizarre tragedy of Lake Nyos, in Cameroon. This beautiful, normally placid lake sits in an old volcanic crater. Carbon dioxide naturally seeps from geothermal sources below the lake and dissolves under pressure in the cold layer of water at the bottom. On August 26, 1986, the deep waters of the lake reached their CO_2 saturation point, and without warning the lake "turned over," its bottom layer shooting to the surface in a violent, frothy eruption of seltzer water that flew some eighty meters into the sky. The CO_2 then formed an enormous cloud that, heavier than air, slid down the mountainside and quietly smothered 1,700 people in the valley below. Today, the lake is constantly vented with a fountain to keep it from happening again.

Despite the risks, carbon sequestration has gained sudden momentum with the U.S. coal industry and the U.S. Department of Energy. It isn't seen as a way to catch the vast stream of CO_2 coming from existing plants (nobody has yet imagined a way to do that), but rather a hoped-for part of a futuristic generation of thoroughly redesigned coal plants decades away.

To have any faith at all in carbon sequestration, though, it appears we would have to set up a tremendous system of international governmental oversight to ensure compliance. Every company and every country would be faced with a strong, ongoing economic incentive simply to vent the CO_2 to the air instead of going to the expense of, say, hauling it to the nearest ocean or salt mine and pumping it into the depths. Since CO_2 mixes so thoroughly in the global environment, cheating this way would be incredibly hard to detect absent near-constant oversight by some regulatory authority. Some American coal industry advocates have argued over the years that the climate scare is driven by international bureaucrats who want to expand their global reach dramatically. Now the industry is promoting a technology that seems to demand precisely that.

Is carbon sequestration a dangerous and desperate effort to cling to an energy technology we are now better off abandoning, or is it a reasonable way to make palatable a still critical source of energy? The answer depends mainly on what other sources of energy we can feasibly turn to.

In most developed countries, the good sites for hydroelectric dams have already been used up, and all new dams have major environmental ramifications. Nuclear fission continues to face its own waste sequestration problem, not to mention heightened safety concerns based not just on the threat of accidents but on the threat of terrorist attacks. Nuclear fusion has raised hopes for years but has yet to materialize. Oil produces its own greenhouse gases, and brings its own well-known political, environmental, and supply problems. Natural gas, a much cleaner source of power than coal, contributes far fewer greenhouse gases and other pollutants for the energy obtained. It is widely seen as the fuel that will help the world bridge the distance between our carbon-rich present and a decarbonized future. But, even though supplies are larger than we used to think, it will be difficult for natural gas alone to fill the gap left by coal.

Wind and solar power represent the ultimate in environmental sustainability, but there is a problem with reliability: The sun doesn't necessarily shine and the wind doesn't necessarily blow exactly when you need it to. Electricity is a product with essentially no shelf-life; it must be generated to meet demand exactly as it arises because we haven't come up with a practical way of storing it. Coal can be handily piled up next to our power plants for use as needed; when a heat wave strikes and all the air conditioners in town go on, we can just shovel the coal in faster and keep ourselves in a state of climate-controlled comfort, even as we increase the risk of the next heat wave.

All in all, trying to find an environmentally safe and reliable substitute for coal can be a depressing proposition. At least, it's depressing if you're thinking only in terms of replacing the coal plants at the hub of our electric grid while keeping the other elements of the energy infrastructure in place.

The picture becomes vastly brighter, though, and even encouraging, if you let yourself think more broadly. Those who do typically envision a world that runs not on carbon but on hydrogen. Hydrogen is the most abundant element in the universe, and it can be made out of water (H_2O) simply by separating it from the oxygen atoms. For this reason, as long ago as 1874, Jules Verne called hydrogen extracted from water "the coal of the future." Sadly, it takes more energy to make hydrogen from water than you actually get out of the hydrogen; otherwise water itself would be a fuel source. Still, hydrogen could become a vitally important way to store and move energy, and combined with modern renewable technologies, it could let us do what our ancestors could not: harvest huge amounts of usable energy from solar income.

Under this scenario, wind and solar power, and in some places geothermal power, would be used to separate enormous amounts of hydrogen from water, and that hydrogen would then be distributed through pipelines or in tanks to wherever it is needed, as oil and gas are today. It could be burned in power plants, and virtually the only emission would be water vapor. Or we might bypass the power plant altogether, and even the act of burning. Fuel cell technology, first developed for the space missions, allows us to turn hydrogen directly into electricity or heat. Not only can fuel cells heat your home and power your appliances, they can also run your car, and the major auto companies are investing heavily in them.

This vision of a hydrogen-fueled economy does not depend on great technological leaps. We have the technology; the problem is cost. Renewable technologies still cost more than the direct costs of existing carbon technologies, and especially of coal, prices for which keep falling as the industry turns from miners to mining machines. Renewables are much cheaper than they used to be, though, and once they break out of energy niches into the mass market, costs are likely to plummet. Those mass markets may be only a few years away—probably not in the United States, where climate policies still face fierce opposition, but in Europe, which is currently taking the lead in adopting policies limiting greenhouse gases.

One key to making the shift to new energy technologies may be to abandon the highly centralized, mass-produced approach to energy that has so long prevailed, epitomized most fully by modern coal plants. Coal-fired power plants, like the steam engines of the past, are more efficient the larger they are. Over the years, many coal plants have been built to mammoth proportions, and the need to include expensive pollution control technologies in the plants just adds to the incentive to build them bigger. Because of this trend, and because electricity has always been controlled by highly regulated monopolies, little room has been left for competition and new ideas; as a result, the technological evolution of the industry has been stunted. Today, our computers are largely electrified by coal-fired technology that has changed little in its basic design since Thomas Edison. Coal, the fuel that once powered an energy revolution, is holding back the next one.

This situation is already changing for other reasons, as governments throughout much of the world (including Europe, China, and the United States) open their electric markets to competition. Power deregulation—if it is artfully combined with new climate policies—could prompt a surge of technological change reminiscent of the emerging steam age, or, more recently, the information age. Not only have information technologies resulted from and spawned tremendous spirals of innovation, but they have also greatly stimulated the global economy. A suddenly competitive energy industry freed from the undermining effect of artificially cheap coal-fired power could do precisely the same thing.

There would surely be social and economic costs surrounding the exit of coal from the world energy scene, and acute pain in areas still heavily dependent on coal production. The industrial revolution showed us how devastating abrupt economic changes can be for people, particularly if no broader societal effort is made to limit the damage. There is, though, reason to believe that the advantages of the shift from coal to less inherently dirty and dangerous energy sources would exceed the costs, quite apart from the climate benefits, because our air would be strikingly cleaner. Acid-damaged ecosystems would be revived; mercury would be less of a threat to our children; and lung-burning and crop-damaging smog would drop dramatically, and when the internal combustion engine is replaced, would vanish. Our long-range vision, which coal burning has impeded both literally and figuratively, would suddenly improve. And we would suffer far less from the widespread asthma attacks, lung disease, and premature deaths that particulates inflict upon us.

In developed nations, tens of thousands of lives yearly might be saved by not burning coal, and in China, perhaps hundreds of thousands of lives would be saved each year (not to mention the thousands of lives saved by closing the mines). In just a few coal-free years, more lives would be saved than the coal jobs that were lost. But stories about such benefits would be unlikely to appear on the evening news because we would not know whose lives had been saved and because the quite genuine plight of the coal regions would provide pictures and stories with much more dramatic appeal.

If the highly concentrated power system that coal represents is indeed replaced by much smaller and dispersed energy sources contributing to a common network, there would be other benefits, too. Contributing sources could be spread throughout society: power flowing from photovoltaic cells built invisibly into the roofs and walls of our buildings, wind turbines spinning over our farmlands, and no doubt other technologies we have yet to imagine. The broad base of such an energy system would make it inherently less vulnerable to disruption by accidents, storms, market manipulators, and terrorists. And, while this is by no means certain, moving from the concentrated power of coal to the more widely dispersed power that nature provides on the surface could prompt a parallel dispersal of political and economic power; we could see a reversal in the trend toward ever-larger corporations in favor of a power distribution that is more democratic and inherently closer to the Jeffersonian ideal.

The transition would take decades, of course. During that time, much of our electricity would come from natural gas plants. Some would also come from the coal fields, but not necessarily in the form of coal itself. One coal-based technology just getting underway is the use of methane—the explosive ingredient in the fire-damp that has killed so many miners—extracted from the coal beds. Today, most of this methane just leaks out or is vented. Not only does this represent the loss of a fuel much cleaner than coal, but it is one more way that coal production destabilizes the climate: Unburned methane is a greenhouse gas twenty-one times more potent than carbon dioxide. By focusing on methane extraction instead of coal, some coal companies and coal regions could play an energy role well into a post-coal era.

With luck, the computer industry's example would apply to energy in one more way: An intensely competitive market that rewards innovation would promote the rapid evolution of energy technologies while the prices quickly drop. In time, the taxes, subsidies, and treaty commitments needed to get this technological ball rolling would all become irrelevant. Carbon-free energy sources would have come into their own in the marketplace. By the time we've figured out how to build that new generation of coal plants from which all the carbon dioxide can be captured and sequestered, there may well be no interest in such a primitive energy technology.

In imagining possible futures, it's important to factor in one last critical energy source: excitement. There was a time when coal was actually fun—not the mining, which was never fun, but the building of a powerful new coal-fired world, which inspired distinct bursts of imagination, enthusiasm, and daring at various historical moments. Matthew Boulton's excitement over his partner James Watt's steam engine came from his belief in it as the key to building a magnificent new society of unlimited possibilities. George Stephenson was surely thrilled to zoom across the English landscape, at speeds never before reached, on railways he knew would transform the nation and the planet. Jacob Cist and his young crew shot the rapids of Pennsylvania's Lehigh River on flimsy arks filled with coal mainly for the glory and adventure of helping create a new industry to fuel a new world. Even Franklin B. Gowen, when he organized the anthracite cartel in the name of empowering a growing nation, was probably having an excellent time.

Today, though, the thrill of coal is largely gone, and it's not likely to come back, even if technological breakthroughs can reduce its environmental impact. Bright, ambitious people eager to change the world will be less inspired by the challenge of building a vast new carbon sequestration infrastructure—essentially a system of perpetual waste management—than by the challenge of building completely new energy industries with no waste at all. Enthusiasm and talent will flow more freely to the new energy industries, not to mention the backing of visionary financiers and environmentally concerned investors. Most people would simply rather direct their passion and money toward creating a new world than toward retrofitting the old one, at least when the old one is as deeply flawed and widely unloved as your average coal plant.

It is, in short, hard to imagine that we'll ever make the sweeping investments needed to make coal use truly sustainable when we appear to have better, and ultimately probably cheaper, options even among the technologies we already know about. But that still leaves open the question of when we will have the political will to abandon the investment we've already made in coal—and, in places like the United States and China, the investments we continue to make.

OVER TIME, PAST GENERATIONS have had deeply conflicting feelings about coal. It's been seen as a mark of poverty and as the epitome of fashion, as an annoying nuisance and as the foundation of human progress, as a sinister substance linked to a demonic underworld and as a destiny-shaping gift from God. Whether we and the people who will follow us remember coal for the way it helped build our civilization or the way it helped undermine it depends—as so many more important things do— on the climate ahead.

Scientists are confident that the climate will continue to warm somewhat, no matter what we do now, as the world responds to the greenhouse gases we've already emitted. That doesn't mean our fate is sealed, though. If in the next few years we move quickly and aggressively toward climate-friendly energy sources—and if we are lucky and the earth is not as sensitive to the greenhouse gas build-up as some computer models suggest—we might see only the lower and much less traumatic end of the projected warming range. We would also greatly reduce the likelihood of pushing the climate past the point that triggers the kind of enormously destructive and abrupt climate changes our planet is prone to. We won't eliminate that risk, though, because nobody knows just where that critical point is.

If we do trigger drastic climate changes, all of coal's contributions to the empowerment of humanity will be overshadowed by the enormous price of that power. Our excuses for continuing to burn coal while ignoring the threat of climate change for so many years—our lack of scientific certainty, our desire to keep our electric rates low, our fear of a slowed economy, and our reluctance to make sacrifices others are not forced to make—will ring hollow to those coping with the catastrophic consequences of our actions.

If, on the other hand, we can actually make the transition to a safer energy system before we cause more than mild climate changes, our coal use won't be strongly condemned by future generations. Some of our descendants may simply see coal as a strangely primitive fuel and wonder how we tolerated it for as long as we did. The more thoughtful among them may recognize it as an important energy source that, for all its faults, brought us through a sort of prolonged industrial childhood and ultimately gave us the power to build a world that no longer needed coal.

ten

Tipping Points:
Epilogue to the New Edition

O<small>NE</small> S<small>ATURDAY</small> <small>MORNING</small> in June 2008—five years after this book was first published—I got my kids up early so we could witness a true spectacle. We joined thousands of others positioned along the Mississippi River to eagerly await the implosion of a local landmark. It was, in fact, the very smokestack I had photographed years earlier for this book, rising above the High Bridge coal plant about a mile from my home in St. Paul, Minnesota.

At close to the appointed time, the explosives placed at the structure's base exploded. The 570-foot-high stack—taller than any building in the city—dropped down a few feet and then seemed to hesitate before beginning its slow and elegant arc to the ground. Several seconds later it landed with a highly satisfying crash, shaking the earth, setting off car alarms, and shooting a thick brown cloud of dust into the air. Cheers rose from the delighted crowd.

I was hoping that this demolition was a sign of things to come, and indeed many old coal plants in the United States have since been or are being closed down. Coal as an industry, though, has refused to go down with anything like the grace of that smokestack. Over the last dozen years, coal interests have pushed to deepen the nation's and the world's coal dependency, even as the science shows that to preserve a livable climate we must slash our coal emissions—and fast. The absurd denial over climate science, which the coal industry pioneered and still perpetuates, has now taken on a life of its own within the U.S. political Right, paralyzing Congress in the face of humanity's biggest challenge. Meanwhile, global coal use and carbon emissions have skyrocketed while the world sets new records for rising temperatures and seas, melting ice and permafrost, acidifying oceans, and the more extreme storms, floods, droughts, heat waves, and wildfires that characterize a destabilized climate.

If we lived in a world where global transformations happened only in a steady, linear fashion, it would be difficult to hold out hope for humanity's chances of transitioning away from coal and other fossil fuels fast enough to preserve a bright future. Fortunately, though, we live in a world where—as shown by the history of coal itself—slow changes can reach a tipping point and suddenly accelerate. And we are making encouraging progress toward favorable tipping points in energy economics, technological performance, social movements, and even public policy (despite some conspicuous failures in the political realm).

Can we reach these world-saving tipping points before we cause disastrous warming or even hit a natural tipping point at which global warming accelerates beyond humanity's ability to influence or adapt to it? The answer depends in no small part on how fast we can replace coal.

IN THE UNITED STATES, before we could address the question of how deeply to cut our nation's coal emissions, we spent years fighting over whether to dramatically raise them by building additional coal plants across the country. The wave of proposed new coal plants came to be known as "the coal rush," the beginning of which I mention in chapter 7. At its height in early 2007, over 150 announced new coal plants were in various stages of development, permitting, or construction. If built, they would have locked in decades of high carbon emissions and diverted hundreds of billions of dollars away from the just-barely-doable challenge of decarbonizing our energy system over the next few decades.

People in the business of building new coal plants were practically giddy over the economic opportunity; I attended one industry conference in 2005 where a speaker foresaw the "best of times" ahead for coal-fired electrical generation. Outside the industry, though, many looked on with incredulity at what seemed to be evidence of a sort of mass delusion: the belief that we could ignore increasingly dire climate warnings even when making our longest-term, most-polluting investments. Among the most alarmed were those working in the nation's scattered environmental nonprofits. After this book was published, I worked for years in coalition with many such groups in Minnesota and the Midwest, mostly on behalf of the Union of Concerned Scientists, a national group active in support of clean energy and climate protection policies.

At times, the sheer economic momentum of the coal rush felt overwhelming. These were often billion-dollar-plus projects, proposed by politically powerful utilities and supported by intimidating legal firepower. But the staggering amount of carbon these new plants would emit for decades was so clearly unacceptable that it galvanized its own resistance movement. Environmental groups that had focused largely on other issues, like conservation or local air and water quality, started to work collectively to stop the coal rush, helped by foundations that understood the scale of the threat and the value of a coordinated response.

To the surprise of many, the resistance won: about four-fifths of the proposed coal plants were blocked or cancelled, and the trend toward project failure was clear even before the 2008 economic crisis cut the legs out from under the remaining proposals. Opponents had successfully raised legal and regulatory barriers, stressing both environmental and economic costs. Ratepayers had objected, especially as estimated construction costs ballooned. And the uncertainty created by the obvious need to regulate carbon in the future gave utility regulators and investors cold feet.

It was a useful lesson in how seemingly unstoppable economic momentum can in fact be stopped and proof that when you are talking about human behavior, even the most widely accepted and well-supported predictions can be quite wrong.

ODDLY ENOUGH, the height of the coal rush in 2007 was also the height of a short-lived golden age of relatively bipartisan U.S. climate action. Public concern over climate, while never as high as the facts warrant, rose in 2006 and 2007. This was the period of maximum media attention to the issue, with broad coverage of Al Gore's 2006 documentary *An Inconvenient Truth* and a major 2007 report of the Intergovernmental Panel on Climate Change (IPCC) announcing that global warming was now "unequivocal."

With so little happening at the federal level during those years, activity shifted to the states. By the end of 2008 more than two dozen states already had renewable energy standards that required electric utilities to get a small but growing share of power from renewable sources, and many others had programs to promote energy efficiency. Eighteen states had set their own long-term greenhouse gas reduction targets (typically aiming for emissions reductions of 80 percent or so by 2050, with specific interim goals), and over thirty states had created carbon-reduction plans, often tasking stakeholders from different sectors of society with sorting through an array of pollution-cutting options.

In 2007 I was invited by Minnesota's Republican governor to join with over fifty others to create a plan to meet the state's carbon-cutting goals. Sitting around the table were people from power companies, industry, agriculture, nonprofits, and many other sectors. The work was wonky, data driven, messy, frustrating, occasionally confrontational—a few of us were simultaneously fighting to stop a new coal plant that others at the table were trying to build—and sometimes inspiring. After months of squinting at graphs and trying to predict long-term future trends, we agreed on a plan, with unanimous support for almost all its elements.

As in other states, our plan included many specific policy actions plus a catchall policy tool called cap-and-trade—a regulatory approach then sweeping the nation. Cap-and-trade lets the government cap the total amount of pollution produced by an entire segment of the economy rather than putting emission limits on each facility the way traditional regulations do. Facilities can then buy or sell the right to pollute, creating a market that drives pollution cuts where they can be most cheaply made. Cap-and-trade has been used to great success to reduce sulfur dioxide from coal plants and has long been hailed as the flexible, business-friendly way to use market forces to cut pollution.

A nationwide cap-and-trade system seemed all but inevitable during this period. Twenty-two states, choosing not to wait for the expected federal system to emerge, started setting up their own regional cap-and-trade markets, with another ten states participating in the process as "observers" in case they wanted to join these regional markets later. In 2008, both presidential candidates supported cap-and-trade. After the election, a major push for the long-awaited federal cap-and-trade legislation was launched. The House climate bill, known as Waxman-Markey, passed a floor vote in 2009.

Some in the coal industry stridently opposed any such bill, like Bob Murray, the outspoken CEO of Murray Energy. In earlier congressional testimony he had denounced attempts to limit carbon emissions as part of a "deceitful, hysterical, out of control, rampage perpetrated by fear-mongers." But other coal interests were well represented in the Waxman-Markey bill negotiations, and they got major concessions, including a river of money that would have helped the coal industry by subsidizing carbon capture and storage (CCS) technology.

Despite many shortcomings, Waxman-Markey would have finally put in place a plausible mechanism to achieve the deep, long-term greenhouse gas emission cuts we so desperately need. But this was not to be the climate policy tipping point that we had long worked toward; the bill was unable to get the supermajority support needed to overcome a threatened Senate filibuster, and cap-and-trade was soon declared dead.

Like the coal rush, yet another seemingly unstoppable trend was stopped. This time, though, it was the wave of climate protection efforts that came crashing down—the victim of a political tsunami that, among other things, would sweep climate denial from the backwaters into the mainstream of the Republican Party.

WHILE CLIMATE DENIAL had already taken root among political conservatives by 2006, they did not yet control the Republican Party and could not stop moderate Republicans from working toward climate solutions. By 2008, however, the party's base was increasingly passionate about promoting fossil fuels, and delegates were chanting "Drill, Baby, Drill" on the national convention floor. In 2009, only eight Republicans voted for the House cap-and-trade bill. By 2012, all but one of the many Republican presidential candidates denied the need to fight global warming. Many of them had to awkwardly recant their former positions on the subject, including Mitt Romney, Newt Gingrich, and Minnesota's former governor Tim Pawlenty, who called his years of support for cap-and-trade a "stupid" mistake (and who had already shelved the emission reduction plan our stakeholder panel crafted for him in 2007).

The coal industry certainly helped stoke this political shift. In 2009, Massey Energy, at the time the largest coal producer in Appalachia, sponsored a West Virginia Labor Day rally attended by 100,000 people; Don Blankenship,* then the company's CEO, assured the crowd that "global warming is pure make-believe." Mixing antigovernment and anticorporate rhetoric with a generous splash of xenophobia, Blankenship warned that cap-and-trade was the creation of "your utility company, as part of corporate America, and the government . . . colluding to trick you so as to tax the poorest and hardest working Americans" and give their money to "the banks," "illegal immigrants," and "perhaps even to the Chinese." The coal industry helped undermine climate action in more covert ways as well; bankruptcy filings by Alpha Natural Resources (which purchased Massey Energy in 2011) revealed that in addition to secretly funding climate denial groups, it funded the work of people involved in the sustained harassment of climate scientists.

Outside coal country, the oil industry—particularly that segment most ideologically opposed to regulation—has been play-

---

*Blankenship, a highly controversial figure in West Virginia, would later be indicted by a federal grand jury for conspiring to routinely violate mine safety standards at the Upper Big Branch coal mine, where an explosion killed twenty-nine miners in 2010.

ing the larger role in undermining political support for climate protection. Charles and David Koch—the billionaire brothers whose fortune is rooted in oil refining and whose extreme antigovernment views were for decades consigned to the political fringe—have for years through their foundations heavily bankrolled groups that dismiss the climate threat and attack climate protection efforts. One group, Americans for Prosperity, held rallies against cap-and-trade in cities across the nation. When the Tea Party protests first flared in 2009, this same group raced in to organize rallies, train activists, and otherwise help make the Tea Party a viable political force while ensuring that climate laws became one target of its anger. The staggering sums the Koch brothers continue to raise and spend to shape U.S. politics have helped keep carbon reduction entirely off the Republican Party's agenda ever since, except as something candidates must vehemently oppose to prove their conservative bona fides.

As the Tea Party took control of the Republican Party, it paralyzed Congress as well as many states, preventing further climate action and even undoing some of the policies put in place years earlier. The regional cap-and-trade systems in the Midwest and West died before they could be launched (though such systems did emerge in the Northeast and California, where they have operated very successfully for years).

The rise of high-decibel climate denial during these years had the desired effect on public opinion too. Between 2006 and 2009, the percentage of Americans telling pollsters there was "solid evidence" of global warming fell from 77 to 57 percent, and it fell across the political spectrum (from 91 to 75 percent for

Democrats, from 79 to 53 percent for Independents, and from 59 to 35 percent for Republicans).

In fact, there is a staggering amount of global warming evidence, and it is so convincing that "virtually every national scientific academy and relevant major scientific organization," in the words of the American Association for the Advancement of Science (AAAS), has issued statements proclaiming that climate change is putting the world at risk. Some 97 percent of individual climate scientists endorse the mainstream view, which was developed over decades of intensive peer review of multiple lines of evidence collected by thousands of scientists. The climate consensus is similar in strength to that linking smoking to lung disease, according to the AAAS. And yet, even today, Americans remain largely unaware of this overwhelming level of agreement; a 2015 poll found that only one in ten respondents accurately understood that over 90 percent of climate scientists have concluded global warming is happening.

The declining public concern over climate change, plus the contrived "Climategate" scandal,* helped doom the international climate talks in Copenhagen in December 2009. Negotiators had worked for years to replace the Kyoto Protocol with a new deal covering both developed and developing countries, but the gathering of the hundred-plus world leaders and 40,000

---

* Hackers stole thousands of e-mails from climate scientists and then found a few phrases to mischaracterize, suggesting a few scientists were altering the data to make warming seem worse than it was. Six separate investigations into these scientists and their data have fully exonerated them of any wrongdoing. The hackers have never been found or prosecuted.

other people at the Copenhagen talks failed to produce such an agreement.

By the end of 2009, the years-long campaign to put in place a legal structure to preserve the climate was in shambles, both domestically and internationally. After a brief drop in climate-changing pollution during the economic crisis, global emissions were about to come roaring back in 2010, returning the world to an emissions trajectory close to what just a few years earlier had been seen as the worst-case scenario.

A FAVORITE SAYING among those trying to move the world beyond fossil fuels is "The Stone Age didn't end because we ran out of stones." It gets a reliable laugh in part because the idea that we might outgrow fossil fuels before we actually use them up is still pretty novel. Of course, such thinking is entirely foreign to coal companies. In 2010 Peabody CEO Greg Boyce assured a congressional committee that with respect to the trillions of tons of coal in the world, "we will use it all"; the only question is how.

The idea of leaving fossil fuels in the ground has gained new urgency, though, as the climate threat becomes clearer. The more the Earth warms in the decades ahead, the more we have to think about the unthinkable, like widespread dust bowl conditions over vast swathes of the planet, devastating global food supplies; mass extinctions and the collapse of both terrestrial and marine ecosystems, with the oceans at particular risk because our carbon emissions not only warm but acidify the water; and rising sea levels that will eventually force the abandonment of the world's coastal cities and low-lying regions.

While we know the seas will continue to rise, there is genuine scientific uncertainty over how much and how fast. Scientists make vastly different predictions depending on their assumptions about the not-well-understood melting dynamics of the vast ice sheets in Greenland and Antarctica. The IPCC says it is "likely" (which in the IPCC's parlance means a probability of 66 percent or more) that by 2100 seas will rise from half a meter up to a meter if we stay on a high emissions path. This is plenty alarming, given that about 150 million people live within a meter of high tide. And yet this projection doesn't claim to capture the high-end risks that might not be likely but are distinctly possible. One new analysis that focuses more explicitly on the upper limit of sea level rise by 2100 estimates a nearly 5 percent chance of it reaching a far more dangerous 1.8 meters or higher.

In the very long term (centuries to millennia), seas could rise several meters more—dramatically redrawing Earth's coastlines and putting many of the world's most populated areas under-water—as a result of our generation's pollution setting off an unstoppable disintegration of the polar ice sheets. Indeed, in 2014 scientists announced that one particularly vulnerable part of the West Antarctic ice sheet had already "reached the point of inevitable collapse." Most studies' assumption that it will take many centuries for the ice sheets to melt has prevented wide-spread alarm over these projections, but one recent and highly controversial bombshell of a study concludes the process could happen much faster: although not yet peer-reviewed, a study by several prominent experts warns of potential sea level rise of *several meters* in fifty, one hundred, or two hundred years—a catastrophic impact unfolding possibly even within our lifetimes.

And the more the Earth warms, the more we risk triggering severe natural feedback effects that would suddenly switch the climate into a different state, such as the release of vast amounts of methane from the permafrost or seafloor. A recent joint report of the British Royal Society and the U.S. National Academy of Sciences says that computer models do not predict such a disaster scenario "in the near future" but adds that because we are moving toward an atmosphere the planet has not experienced in millions of years, "we are headed for unknown territory," and such disasters "cannot be ruled out."

To reduce the dangers we would face from a severely destabilized climate, in 2010 delegates from virtually every nation in the world formally adopted the goal of limiting global warming to no more than 2 degrees Celsius above preindustrial temperatures (while further considering a limit of 1.5 degrees). Many scientists argue that allowing two degrees warming—more than double what we have already seen—is itself highly dangerous and that we should in no way resign ourselves to it. But what lies beyond two degrees warming is unquestionably even more dangerous and to be avoided at all costs.

Achieving even the two degree target will take immediate and sustained action on a global scale. Scientists have calculated how much more carbon humanity can emit on top of our past emissions (which have piled up in the atmosphere over the centuries) and still reasonably hope to stay within two degrees of warming. This spells out the world's "carbon budget," and right now we are on a path to blow through that budget entirely by 2040. This is according to an estimate made by the International Energy Agency, and it already factors in the emission reduction pledges made by many countries by mid-2015 and makes very optimistic assumptions about putting carbon capture and storage on coal plants after 2025.

It is urgent, therefore, that we hit our global emissions peak as soon as possible and then reduce emissions as fast as possible. Doing so will help us stretch out our carbon budget, pushing that 2040 date back a bit and giving us time to take a more plausible path to zero net carbon emissions. At this point, though, it will need to be a steep downward path because we have delayed real action for so long.

What does the two degree warming limit mean for coal? It means leaving the vast majority of it in the ground, along with most of our other fossil fuel reserves. *Unburnable Carbon,* an influential study by the Carbon Tracker Initiative, a London-based financial think tank, took a new approach to the problem in 2011. The authors back-calculated how much of the world's fossil fuels can be burned and still leave us with a reasonable chance of staying within two degrees of warming. They found that we can afford to burn only about 20 percent of known reserves. (A later study out of the University of London would find that 92 to 95 percent of known U.S. coal reserves should stay buried if we want a better-than-even chance of keeping warming under two degrees.) The Carbon Tracker Initiative warned that the presence of so many effectively unburnable fuel reserves listed as assets on the books of oil, coal, and gas companies around the world artificially inflates the value of those companies and creates an enormous financial bubble waiting to burst. By putting the issue in such stark financial terms, the *Unburnable Carbon* report helped raise the alarm among investors.

The report also drew the attention of the burgeoning new climate protection movement. Keeping carbon buried has become a core tenet of the global grassroots climate group 350.org, founded by author and activist Bill McKibben (who did much to publicize the concept of unburnable carbon). On the coal front, instead of just trying to close coal plants, climate activists started focusing in the last few years on preventing the expansion of coal mines, railways, and ports. The eyes of the world—including those of investors—shifted from the carbon dioxide coming out of the smokestacks to the coal coming out of the mines.

DESPITE THE PROFOUND political setbacks in the U.S. Congress and international climate negotiations, there have been distinctly encouraging signs of progress in other arenas. We may be nearing some crucial tipping points related to energy economics, technology, and investor sentiment.

While global emissions rose steeply through 2013, the carbon emissions from the U.S. power sector actually peaked in 2007 and have since dropped about 15 percent. This is due to a shift in recent years from coal power to natural gas and renewables.

(Total U.S. greenhouse gas emissions are down about 9 percent since peaking in 2007.) Instead of trying to predict how many new coal plants will be built, analysts are now trying to predict how many existing ones will be closed. Instead of providing half of U.S. power, as it did in 2003, coal provided only 39 percent in 2014—and coal's market share is still falling.

The economic outlook of the U.S. coal industry has fallen even more strikingly, as what observers are calling a coal crash has replaced the coal rush of the last decade. In 2010, Peabody's CEO predicted we were in "the early stages of a long-term supercycle for coal" and his company especially would benefit. The next year Peabody's stock did rise, to over $70 per share, but it quickly began to tank, falling to lows of about $1 per share in 2015. Many of the other top coal producers have experienced similarly dizzying plunges in their share value as investors flee the coal sector and bankruptcies mount (including the bankruptcy of Alpha Natural Resources, one of the nation's largest coal producers). The ever-quotable coal baron Bob Murray says we are seeing "the absolute destruction of the United States coal industry. It isn't coming back." He blames "environmental alarmists," "liberal elitists," and a war on coal waged by "the insane, regal administration of King Obama."

Even before its current economic nosedive, the coal industry was complaining about President Barack Obama's "war on coal," as if coal's troubles were the result of one man's inexplicable personal vendetta against the substance itself. This "war on coal" rhetoric is just the newest example of humanity's centuries-long tendency to anthropomorphize coal. But in the past—whether coal was seen as a holy and civilizing force, as a demonic substance, or as mighty King Coal—the focus was on coal's power, for good or for ill. Now the industry is cloaking coal in the mantle of victimhood, characterizing it as a weak entity suffering unfair harm by larger powers. It is a risky strategy; despite a general sympathy for unemployed miners, most people have no particular sympathy for the coal industry. Indeed, polls show that regulation of coal plants, including their carbon emissions, has broad public support.

In truth, much of coal's decline can be attributed to market forces. The U.S. natural gas industry has been transformed over the last few years because of a controversial new drilling technique called hydraulic fracturing, or "fracking." Its use has boosted gas production and lowered gas prices, making it harder for coal to compete in the power markets.

And yet environmental regulations have played a major role in actually closing coal plants. These include, among others, rules to cut the plants' particulate emissions, which still kill thousands yearly, and to cut the mercury emissions that still threaten fetal and infant brain development. The Environmental Protection Agency (EPA) also began to take a stricter approach to mountaintop-removal mining, which by 2009 had blasted over five hundred mountains in Appalachia. In many cases the EPA had no choice but to regulate; suits brought by environmental groups and others led to court orders requiring the EPA to meet its obligations under existing environmental statutes.

This happened in the case of carbon too. In 2007, the U.S. Supreme Court ruled that the Clean Air Act required the EPA to regulate carbon dioxide if it found (as it later did) that CO_2 endangered public health or welfare. When Congress failed to act on climate change, the EPA was not just authorized but obliged to pick up the slack. In the summer of 2015, the EPA finalized rules under which new coal plants would have to use costly carbon capture and storage technology. At the same time, it put in place the first-ever federal limits on the carbon emissions of *existing* coal plants. Nearly a quarter century after the Earth Summit, when the United States and the rest of the world signed a treaty pledging to protect the climate, the United States is finally regulating its biggest greenhouse gas polluters (though it will take a few more years before the regulations actually cut emissions).

The EPA's climate rule for existing plants, called the Clean Power Plan, is the biggest step the nation has ever taken to protect the climate. The rule aims to reduce U.S. power-sector emissions by 32 percent by 2030 (as compared to 2005 levels, and we are nearly halfway there already). States have unprecedented flexibility in choosing their approach. For example, at the coal plant level they could reduce emissions a little by improving plant efficiency or a lot by adding carbon capture and storage (both expensive options). Or they could reduce emissions at the grid level by replacing coal power with more power from natural gas, renewables, or nuclear plants or by cutting consumption through energy-efficiency programs.

This new climate rule is proving the last straw for the oldest and dirtiest coal plants, which face new costs under other EPA rules as well. Adding the climate rule to that mix has intensified the economic pressure to just retire these plants rather than try to retrofit them. And where necessary, the anti-coal movement that rose up during the coal rush has been pushing coal plants toward closure. That movement is now more organized and much better funded; notably, Sierra Club's Beyond Coal campaign has garnered $80 million from Michael Bloomberg's foundation, though many other local and national groups and foundations have been part of the effort to close coal plants.

It was always clear that the Clean Power Plan would provoke a fierce political showdown with coal's advocates. Even before the climate rules were final, the coal industry and many states sued to stop them, and Kentucky senator Mitch McConnell urged the nation's governors to simply refuse to develop plans under the new law. In that case, though, those states will just become subject to a plan written by the EPA rather than those they craft for themselves, inviting precisely the kind of federal control they most object to.

The U.S. Energy Information Administration predicts that under the Clean Power Plan, the nation's fleet of coal plants will shrink by about a third by 2030, with the remaining plants still providing about a quarter of U.S. power. Then again, this is the same computer model that predicted during the coal rush that the coal fleet would expand by half by 2030. It has a long history of underestimating the costs of coal power and the potential of clean energy. And like all economic models, it is blind to hard-to-quantify things like social and political movements, changes in public sentiment and policy, technological and economic advances, and most of the other things the world depends on to avoid climate catastrophe.

THE U.S. SHIFT away from coal is significantly affecting jobs. On the national level, it has actually increased employment, already creating far more jobs in the wind, solar, and natural gas industries than have been lost in coal; indeed, the solar industry alone employs nearly twice as many people as coal mines do, and it is growing at a blazing pace. But that net employment gain doesn't necessarily help the coal miners or the families and communities that depend on them. The gradual growth in mine employment between 2000 and 2011 has ended, and employment numbers have returned to the steep downward modern-era slide that started in the late 1970s. And job losses will surely continue, especially in central Appalachia where the coal seams are being depleted. The economic pain is most acute in eastern Kentucky and southern West Virginia, which between them have lost over 14,000 jobs since 2011. The Obama administration's 2016 budget includes billions of dollars over the next few years to help coal miners and their communities adapt to the decline of the industry, and local governments are urging their reluctant members of Congress to approve the funding.

Coal's ongoing decline as an industry and employer seems all but inevitable given the dire threat it poses to our climate. The only conceivable way the world can keep coal as a major energy source without courting disaster is through the widespread deployment of carbon capture and storage technology (discussed under its earlier name, "carbon sequestration," in chapter 9). Peabody's CEO Boyce was referring to this technology when he made the counterintuitive claim before a congressional committee in 2010 that "black is the new green," trying to make the case that coal, our highest-carbon energy source, is the path to a low-carbon future.

Between about 2005 and 2010, the IPCC and many others increasingly embraced CCS technology as a viable and potentially crucial means of cutting emissions. The dangerous idea of pumping the captured CO_2 into the deep oceans, originally favored by CCS proponents, was largely forgotten. Instead, researchers hit upon the safer option of pumping the gas into deep saline aquifers, which are widely located around the world.

Many governments—including the United States, the European Union, Australia, and China—were promoting CCS research and trying to launch small pilot or larger demonstration projects. Despite the skepticism I expressed about the technology in chapter 9, in 2008 I coauthored a report for the Union of Concerned Scientists supporting a handful of CCS demonstration projects to test the technology's potential. Only a sliver of the 150-plus new coal plants announced during the coal rush even considered employing CCS, which was not yet used on a commercial scale at any power plant. The problem—in addition to the uncertainty that surrounds any new technology—was cost: adding CCS was predicted to increase the cost of energy from coal plants by some three-quarters.

It was hoped that CCS costs would fall in a few years as the technology was deployed, but those hopes have since dimmed. Ironically, the coal industry may have undermined its own future by opposing climate legislation. The Waxman-Markey bill would have incentivized CCS both by directly subsidizing its use and by putting a price on carbon, and its failure was a severe blow to the development of CCS, at least in the United States. Many announced CCS projects were scrapped, including the troubled U.S. flagship project called FutureGen, which failed to attract enough private investment to get off the ground despite a couple hundred million dollars of federal support.

The world's first commercial-scale coal-fired power plant to use CCS technology, a retrofit Saskatchewan plant, just began operating in 2014. Mississippi Power is constructing a second, much larger coal plant with CCS, but it is years behind schedule and costing nearly triple the original estimates. CCS technology has also proven disappointing in other nations. Ultimately, we cannot look to this technology to play a significant role in climate protection for many more years, and coal with CCS may never prove cost-effective compared to inherently cleaner energy options. In my view, it is still worth trying to develop CCS technology to reduce emissions from the huge number of brand-new coal plants in China and elsewhere, but it looks increasingly unlikely that CCS will be a game-changing technology that will prevent a dramatic decline in coal use. Indeed, even with optimistic assumptions about the future development of CCS, analysts have found that the technology would only marginally increase the amount of coal we could burn before 2050 and still limit warming to two degrees.

Peabody publicly pins its hopes on CCS, supports demonstration projects, and calls for additional government subsidy of the technology; yet one has to wonder if its heart is really in it. The company is simultaneously arguing in state and federal proceedings that climate change is a "non-existent harm." Indeed, it argues that emissions of CO_2, a "benign gas," are good for us. An economist appearing for Peabody recently testified that the benefits of CO_2 emissions exceed the alleged costs "by orders of magnitude" and that "there is no limit for the foreseeable future to these benefits as CO_2 emissions increase."

Usually major corporations trying to resist regulation avoid making such patently ridiculous arguments. Indeed, in chapter 7, I explained that this pro-pollution view was articulated only on the industry's fringe, especially by Fred Palmer, former head of Western Fuels (an opposing party in the 1990s proceeding that first drew my attention to coal). However, Palmer subsequently became the head of Peabody's government relations, and either his view of CO_2 pollution as something to celebrate has spread or else the embattled company figures that at this point it has nothing to lose.

THE DISAPPOINTING PROGRESS of CCS over the years stands in stark contrast to the progress made by renewable energy, which has surpassed all serious expectations. In 2000, world energy experts predicted wind power would double in a decade; in fact it grew nearly sixfold by 2010 and tenfold by 2014. The same experts predicted in 2000 that solar photovoltaic (PV) capacity would reach no more than 3.4 gigawatts by 2010; in fact, the world had installed eleven times that much solar PV by 2010 and more than fifty times that amount by 2014. Together wind and solar still provide only 5 percent of the world's electricity output (the vast majority from wind), but given their rate of growth and their vast potential, energy industry observers predict they will provide 30 percent of global power by 2040 even without additional policy support. And the flow of money is shifting: almost half of the new electric capacity built globally in 2014 was renewable.

Partly, this is because the costs of wind and solar have dropped so precipitously. The price of solar panels fell by 99 percent between 1977 and 2013. While the price of the panels seems to have stabilized, the cost of installing solar systems in the United States keeps falling as installation companies cut other business costs. The cost of energy from new wind turbines is still falling too—down 58 percent just between 2009 and 2014—as the technology keeps improving. In many locations, these clean power sources have already reached "grid parity"—the point at which they can compete economically with conventional power—and their prices are projected to keep dropping.

The striking advances made by these potentially world-saving technologies stem from a healthy blend of private-sector innovation and smart public policies around the world, including the renewable energy standards that most U.S. states have. Federal tax credits have also helped wind and solar compete against coal and other more entrenched technologies (it has proven politically far easier to subsidize clean power than to put a direct cost on polluting power). Obviously, wind and solar are intermittent energy sources, but experience shows we can afford to integrate far more of these resources into our energy grids than most systems currently do.

Eventually, widespread dependence on wind and solar power would indeed threaten grid reliability unless we also widely deploy energy-storage technologies. Fortunately, battery technology costs have fallen dramatically in recent years, and sales are soaring. To cite the most conspicuous example of the industry's ramp-up, Tesla Motors has under construction in Nevada what it calls a "gigafactory"—a mammoth facility to make batteries both for Tesla's electric cars and for electric storage at the utility, business, and even residential scales. This single factory is expected to produce in a year more lithium ion batteries than were manufactured worldwide in 2013 and to drive battery costs down even further. When they reach grid parity, the electricity storage market becomes, in the words of Tesla's CEO Elon Musk, "staggeringly gigantic."

In many ways, solar plus storage is the holy grail of coal replacement. After all, coal is just nature's own ancient and messy battery, storing solar energy delivered eons ago and channeled through an almost inconceivably long supply chain that includes primeval forests, geologic time, coal mines, coal plants, and power grids. Solar plus storage elegantly short-circuits that supply chain, letting consumers both catch and store solar power fresh from the sky. This kind of widely distributed generation offers much less concentrated power than coal offers, but that is not necessarily a disadvantage since our power demand is also widely distributed.

Solar plus storage threatens not just coal but the power sector generally. Utilities can use batteries to integrate more renewable energy onto the grid (and many are starting to), but customers with solar plus storage can get off the grid entirely or just use it for backup power. If enough people do this, the business model of the power sector—originally shaped around centralized coal generation—will be shattered, and utilities are taking notice. So are investors: in 2014, Barclays stunned the U.S. power sector when it downgraded their bonds, citing the threat from rooftop solar with batteries. In short, solar power and battery technologies are evolving so fast they already threaten one of the biggest, most entrenched industries in the nation.*

Of course, renewable energy is not coal's only low-carbon competitor. Nuclear plants have a long track record of providing carbon-free energy (generating 19 percent of power in the United States and 11 percent globally). On the other hand, nuclear power is still plagued with unresolved waste issues, national security concerns, and the potential for devastating accidents, as the Fukushima disaster reminded the world in 2011. It is also hampered by very high costs that—unlike newer, modular, mass-producible technologies like solar—have risen over

---

* The improvements in battery technology also seem to have reduced interest in hydrogen and fuel cells—an energy option I discussed in chapter 9. If electricity can be stored cost-effectively in batteries, there is little reason to convert it into hydrogen and then convert it back into electricity using fuel cells (whether to power cars or the electric grid). While the hydrogen/fuel cell vision of the future continues to have its supporters and investors, there has been only limited progress in turning that vision into reality. Moreover, most hydrogen today is made not the carbon-free way (using renewable power to get hydrogen from water) but from natural gas.

the years instead of fallen. This trend holds true even in a place like France, which is known for its widespread support and deployment of nuclear power. Although dozens of new reactors are under construction worldwide, the International Energy Agency projects that nuclear's share of global power will rise only modestly, to 17 percent by 2050, even in a world with policies in place to keep warming under two degrees. Of course, we could end up relying much more on nuclear power if some of the more radical redesigns currently being pursued—which aim to greatly reduce risk and waste issues—pan out. Either way, nuclear power will have to compete with rapidly improving renewable energy and energy-storage technologies.

Coal's biggest competitor today in the United States is actually natural gas. Gas plants emit about half the CO_2 of coal plants, and the shift from coal to gas is the primary driver of falling carbon emissions in the United States. But there are enormous problems with relying too heavily on gas. When you factor in the methane that leaks from gas production, gas's climate advantage over coal shrinks and could even disappear; leakage estimates vary dramatically, but there is alarming evidence that the EPA's official estimates of methane leakage may be way too low. Also, while the EPA has not found widespread harm to drinking water from fracking, it has found some impacts to both underground and surface water supplies. Moreover, fracking operations sometimes inject the wastewater they produce back underground, a process linked in Oklahoma to a sudden six-hundred-fold increase in earthquakes there.

Another problem is that low natural gas prices don't just knock coal out of the market but also delay the deployment of renewable and nuclear power, meaning that even if methane leaks are controlled, natural gas could end up harming the climate more than helping it. Regulations limiting methane leakage and other environmental harms from natural gas are starting to emerge, and they will probably drive up the fuel's cost somewhat, but in the presence of carbon limits, that cost change will not likely benefit coal plants.

Two other points need to be made in this discussion of alternatives to coal. First, virtually every effort to chart a path toward a low-carbon economy relies heavily on improving the energy efficiency of our homes and businesses. Decades' worth of work has shown that increasing efficiency is almost always the cheapest way to cut carbon emissions. Many efficiency improvements pay for themselves very quickly by cutting energy bills, and there's no question that efficiency advances will play a big part in the decarbonization of the world's economy.

Second, we need not and cannot wait for additional technological breakthroughs before aggressively reducing our carbon emissions and coal dependence. While new breakthroughs are obviously possible and welcome (and research funding should definitely be ramped up), dramatically new technologies rarely if ever emerge from the laboratory fully formed and ready to save the world. After the "eureka" moment, they need to go through years of actual deployment and use in the field, which for the best technologies will lead to a series of incremental improvements and economies of scale that greatly reduce cost. Fortunately, technologies like wind, solar, and batteries have been going down this road for years now, yielding the hard-earned progress in both cost and performance discussed above. And that progress is changing the way the world looks at coal, even in the country currently most dependent on it.

CHINA'S ROLE IN the story of coal and the fate of Earth's climate has only become more dominant since this book was first published in 2003. China's coal use has risen at an astonishing pace, roughly tripling between 2000 and 2013. As a result, coal is now the source of more carbon pollution on a global basis than oil is. Roughly one out of every two tons of coal burned in the world is now burned in China, and the country is now the world's top greenhouse gas polluter by far (though the United States remains the top polluter if you count historic emissions, which have built up in the atmosphere and continue to warm the world today). Coal still supplies a whopping two-thirds of China's total energy use (a measure that includes not just electricity but transportation, industry, and the rest of the economy), compared to 18 percent of U.S. total energy use.

China's skyrocketing coal consumption drew the attention of coal producers everywhere, including the U.S. coal industry. However, selling American coal to China would require building more West Coast export terminals, which would not be cheap or easy. In fact, the industry found itself blocked there by serious local opposition, especially in Oregon and Washington.

And then Chinese officials started talking about what to some had been unthinkable and to others inevitable: limiting China's coal consumption. This was partly due to the appalling air pollution that China's heavy coal diet was causing. One recent Western study found that air pollution kills an estimated 1.6 million people yearly (over 4,000 deaths per day) in China, playing a role in one in every six deaths there. While these fatalities are not all due to coal pollution, coal is certainly the dominant contributor. U.S. coal producers might have thought that China offered an escape from the U.S. "war on coal," but by 2014 the Chinese government had launched what it called a "war on air pollution," a major China-style reform campaign focusing largely on coal.

The choking air pollution in China has provoked the beginnings of an environmental movement there and a general awakening to the health threats of the country's industrialization. Beijing clearly has mixed feelings about this. Greater public focus on pollution supports the official pollution-reduction campaign; on the other hand, public anger threatens political stability. This ambivalence showed in how officials handled *Under the Dome*, a 2015 documentary by a Chinese journalist about the nation's air pollution. The Chinese environment minister initially praised the film, calling it China's "Silent Spring" moment, referring to the classic book by Rachel Carson that is often credited with sparking the modern U.S. environmental movement. But a full-fledged movement was clearly more than Beijing wanted; after being viewed by more than 200 million Chinese online, the viral video was suddenly banned.*

China's war on pollution did not necessarily mean China would reduce its global warming emissions. China could have limited its efforts to putting conventional pollution controls on its coal plants and cutting the pollutants that are killing its citizens but letting carbon emissions continue to rise, as the United States and other developed nations did for decades. Fortunately, China has decided not to follow that dangerous path.

In November 2014, following months of negotiations, President Obama visited China and came back with a landmark deal that many seasoned observers had considered impossible. Under this agreement, the United States pledged to cut emissions 26 to 28 percent by 2025, essentially accelerating the decarbonization process that is already underway. In exchange, China pledged to get its greenhouse gas emissions to peak by 2030 or sooner—a dramatic and essential departure from the trajectory of recent years. China also agreed to get 20 percent of its power from carbon-free sources by 2030. That commitment alone will

---

*On a much tinier scale, I have experienced censorship myself in China. This book was published there in 2004. I was surprised because, while I consider it sympathetic to China's very real development challenges, I did not think officials would have approved its criticisms of China's heavy coal dependence or of Mao. However, I was not able to obtain a copy of the book until my daughter spent time studying at a Chinese university in 2013 and bought one for me. That's when I learned that in the Chinese edition, the entire chapter about China had simply been expunged.

require China to add as much new renewable or nuclear generating capacity in the next fifteen years as its entire current coal
capacity and almost as much as the total generating capacity of
the United States today.

A few days after the landmark deal, China announced that
its coal use would peak by 2020, and there is some speculation—
including among researchers at the Chinese agency that over-
sees economic planning—that coal use there might have already
peaked. While estimates vary, Chinese coal consumption in
2014 seems to have stayed relatively flat or actually fallen, in
stark contrast to recent years, and preliminary numbers show
coal use falling significantly in 2015. Economic data from China
can be notoriously unreliable; indeed, in 2015 officials significantly increased their earlier estimates of the nation's coal use
since 2000. But there is still reason to hope China's apparent
drop in coal use is real, given that it is entirely consistent with an
announced shift in the nation's economic direction.

China has recently decided to fundamentally restructure its
economy to get off the environmentally and economically unsus-
tainable growth track it has been racing along for years. This
means, among other things, shifting away from energy-intensive
manufacturing and purposely slowing economic growth. Of
course, if Beijing decides its economy is slowing too much, it
will face internal pressure to back off its reforms, and coal use
could rise again before the announced peak of 2020. On the
other hand, China might decide to stimulate a slowing economy
by accelerating its already impressive investments in low-carbon
energy.

China is investing aggressively in renewables, with more wind power capacity installed than any other nation, and it recently became the world's largest market for solar panels. It also leads the world in new nuclear reactors under construction. China is trying to increase its natural gas production using fracking, though whether and how much that reduces its greenhouse gases will depend on the rate of methane leakage. And China is about to launch a nationwide cap-and-trade regulatory system, which will be the world's largest carbon market by far, and is considering a new carbon tax.

It is worth pausing to appreciate the layers of irony here. China—an ostensibly Communist country whose economy is still largely controlled by a centralized government—is stimulating market forces to spur innovation and help solve global warming. Meanwhile, in the United States, those who claim to have the highest regard for what market forces can achieve—like the Koch brothers and Republican leaders in Congress—successfully sabotaged legislation that would have let market forces play the same role here. This required the EPA to regulate carbon under the Clean Air Act, so it adopted the Clean Power Plan, though instead of imposing government-set limits on power plants, the EPA invites states to use market forces to cut emissions through cap-and-trade if they prefer. Even so, U.S. opponents are deriding the EPA's rule as akin to communism.

THE CLIMATE DEAL between the United States and China is particularly important because it is expected to spur greater pledges of action by other countries at the international climate

talks in Paris in December 2015. International climate negotiators have been working toward this meeting since the collapse of
the Copenhagen talks in 2009. The years of delay have made the
challenge of avoiding severe climate consequences much harder,
but not nearly as hard as if we miss this opportunity to change
direction.

A "top-down" treaty with binding emission limits on each
nation has proven over the years to be virtually impossible to
negotiate. So negotiators are taking a new approach in Paris.
Instead of trying yet again to agree on limits for each nation, they
are using more of a potluck strategy whereby each nation comes
forward with an individualized plan to reduce its emissions. In
diplomacy-speak, each nation's pledge is called its intended
nationally determined contribution (INDC), and many such
contributions have already been announced.

Another reason for this approach is that everybody knows
the current U.S. Senate would never approve a climate treaty.
When Kentucky senator McConnell became majority leader in
2014, he promised to "go to war" with Obama over coal, and he
is taking that war into the international arena. He called the U.S.
INDC, which relies heavily on the EPA's coal plant climate rule
for its emissions reductions, illegal and sure to fail, and he
warned our international partners that they should "proceed
with caution before entering into a binding, unattainable deal."
He even sent his staff to visit foreign officials to tell them the
White House cannot deliver on its promises. Senator
McConnell and coal's other political champions are doing their
best to sabotage this vital, fragile, possibly last-chance global

endeavor to avoid disastrous climate changes, preferring instead to pretend that the climate threat does not exist.

As with any potluck, some offerings brought to Paris will be unimpressive and a little mysterious. Russia's INDC, for example, is so ambiguously worded that no one is quite sure what it means; it seems to cut emissions only slightly by 2030 and is peppered with loopholes (though it is worth noting that emissions there did already plummet in the 1990s with the collapse of Soviet industry). On the other hand, some offerings will be ambitious: the European Union, long the world's leader in climate protection, has already cut its emissions by 20 percent since 1990, and it's pledging to cut another 20 percent by 2030. The fact that Europe has cut emissions for so long without the guarantee of quid pro quo reductions by other nations is itself evidence that the potluck approach might just work, with nations motivated by a mix of global concern and enlightened self-interest.

What impact will decarbonizing have on the global economy? Analysts at the IPCC, the International Energy Agency, and elsewhere have been studying various computer-modeled scenarios of ways we could meet our energy needs while limiting warming to two degrees Celsius, and they've found scenarios with remarkably low global price tags given the climate benefits. Now a new study by Citigroup, one of the world's largest financial institutions, goes a step further. In what its authors call a "surprising" finding, it calculates that the twenty-five-year financial cost of taking a lower carbon path ($190 trillion) is actually less than the twenty-five-year cost of our current fossil-heavy path ($192 trillion), and that is not even counting the trillions in avoided climate change costs. The savings are largely thanks to the rapidly falling price of renewables and to avoided fuel costs.

To have the modeled scenarios come in at low or no cost is great news, because while ambitious, these pathways are probably not ambitious enough. They are based on IPCC scenarios that typically keep warming under two degrees only by assuming the world can actually achieve net negative emissions globally in the decades after 2050. That is, they assume that the world will either invent technology that affordably pulls tremendous amounts of carbon out of the air or that it will widely deploy carbon capture and storage—not on coal plants, because that would at best get you to zero emissions, but on power plants that burn wood and other biomass. As trees grow, they draw carbon dioxide out of the air; burn enough of them, capture the CO_2 that comes out the smokestack, compress the gas and pump it underground, and you have power plants with negative emissions (assuming the plants then grow back again at a sustainable pace).

This is not unlike the process of coalification through which CO_2 captured by ancient forests was buried by nature, except we would have to do it ourselves on a hugely accelerated basis and vast scale. To avoid burdening our grandchildren with such a Herculean task—and because two degrees of warming could itself prove extremely dangerous—a far saner approach would be to increase our ambition beyond what these modeled scenarios reflect, peaking emissions now and driving them down as fast as possible.

But what is possible? Twenty years ago, it would have seemed impossible that billions of people would by now be connected to the Internet using not just computers in their homes but cell phones in their pockets. Nobody could have predicted that even many of the global poor would obtain digital technology and do so before other basic needs were met; yet by 2013 more people had access to mobile phones than to toilets.

And the stunning technological, economic, and social transformations of the digital revolution were not spurred on by the need to avoid global disaster. Add such an unprecedented motivation to the mix, and the prospect of an energy revolution—demanding its own set of technological, economic, and social transformations—doesn't seem that far-fetched. And indeed, given the remarkable advances of clean energy technologies, we can say that revolution has already started.

THERE IS SOMETHING more than a little surreal about doing precise and detailed cost-benefit analyses regarding a potential global catastrophe. And yet economic models of various kinds have long been at the heart of climate and energy policy debates and even questions over whether to build or close a specific coal plant. The anti-coal and climate-protection movements have had to learn to put their arguments in economic terms, and this has formed a big part of my own advocacy work, starting with the 1990s climate proceeding I discuss in chapter 1.

Economic models can obviously be useful to compare alternate paths forward, but they have huge downsides. They depend on a raft of cost assumptions and other inputs that get obscured or forgotten. They overvalue factors that can be easily monetized and feed a dangerous overconfidence in our ability to know the future. They have a psychologically anesthetizing effect, numbing the sense of urgency we should be feeling given the scale of the threat. They are particularly bad at predicting tipping points, meaning they underestimate both the climate risks and the solutions we might devise. And they introduce into humanity's most consequential long-term choices the narrow, bloodless, amoral values of the marketplace.

In the last few years, though, the moral dimensions of climate change have risen to center stage, along with not a little moral outrage. Climate activists are pushing universities, churches, pension funds, foundations, and others to divest from fossil fuels or at least coal on both financial and moral grounds; more than two hundred institutions have already agreed to do so, including the state of California's public pension funds. Divestment will not bankrupt the fossil fuel companies, concedes one of the movement's leaders, Bill McKibben, "but we can help politically bankrupt them. We can impair their ability to dominate our political life."

President Obama has also increasingly stressed our "moral obligation" to fight global warming, making it a centerpiece of his second term. In June 2015, the pope issued his high-profile encyclical asserting humanity's moral responsibility to preserve the climate (and the need to reduce the use of polluting fossil fuels, "especially coal," without delay), and he repeated this call in his September visit to the United States. Other religious leaders have similarly called for climate action. And most Americans already see the issue in moral terms; one poll found that 66 percent of respondents thought world leaders have a moral obligation to reduce CO_2 emissions, even though only 63 percent were sure humans were causing global warming.

The coal industry, surely uncomfortable in its implied role as the villain in this moral struggle, is trying to recast coal in more altruistic terms. Peabody has launched a new campaign to raise awareness about the humanitarian crisis of global energy poverty and how coal is the only way to address it. The industry is now setting its sights less on China and more on other less developed nations. Bob Murray of Murray Energy has said, "The only way that the folks in India are going to get a light bulb is through coal. It won't be from wind. It won't be solar. It won't be nuclear. I think this is evil. I think it's cruel in the worst possible terms to deny the most poverty stricken people on this planet a light bulb, but that's Barack Obama and his supporters."

However, India is not being prevented from burning coal; on the contrary, it is experiencing a large new wave of coal plant construction. The nation is heavily dependent on coal already and paying a huge price for it: New Delhi now has the most polluted air of any city in the world, and coal smoke has already been linked to the deaths of over 100,000 Indians yearly. An anticoal movement has arisen there in response to growing coal mining and burning, complete with large protests over specific projects (protests that are sometimes violently suppressed), but it remains to be seen whether the Indian coal rush will fizzle as it did in the United States or rocket upward, as it did in China.

The same can be said for the coal rushes in many other developing nations, including Indonesia, Philippines, Thailand, and Malaysia. In short, coal use may be falling in the United States and most of the developed world and may even be close to peaking in China, but on a global basis there is still no end in sight to the expanding role and rising threat posed by coal.

Persuading (and helping) the nations now poised to deepen their dependence on coal to instead leapfrog past it and embrace clean energy is critical to the struggle to preserve a livable climate. And it is yet another reason why the striking advances in clean energy technology—especially wind, solar, and batteries—represent such tremendously good news for humanity. Already, renewables can often bring electricity to those living without it more cheaply than coal can; renewable power can be built in rural areas, which is where the overwhelming majority of people without electricity live, letting them bypass the costly centralized energy grids that coal plants require. And the advantage of renewables over coal will only grow as the costs of renewable technologies continue to fall. The coal industry is half right—global energy poverty is a huge humanitarian problem. It's just that coal is no longer the solution.

ENDING THE AGE of coal before it does too much more to undermine the civilization it helped empower is daunting, but there has been real progress along many fronts. In many developed nations, the shift away from coal is already pushing carbon emissions down. Helpful government policies are spreading, even in the United States where climate denial has blocked congressional action. China seems to be finally veering off its dangerous trajectory, and while many other developing nations are planning to meet their growing energy needs by ramping up their coal burning, the phenomenal declines in the cost of renewables and energy-storage technologies mean they now have alternatives to coal that are often both cleaner and cheaper. We need to greatly increase investments in carbon-reducing technology, but changing energy economics mean much of that investment will pay for itself just in fuel savings. Moreover, investors are increasingly aware that there are tremendous profits to be had in technologies that can meet our energy needs without causing global disaster. And while we still need stronger climate policies, the growing climate movement and the deepening recognition of the moral dimensions of global warming are making it more likely that we can put those policies in place.

At some point, the collective momentum of all these changes will surely overpower the momentum that coal has built up over the centuries. When that tipping point is reached, it will bring about a cascade of additional technological, economic, and social changes that will finally sweep coal from the global scene. The only question—and it is a critical one—is when.

Notes

Chapter One: A Portable Climate

Page 1. Nobles attending Parliament lead demonstrations, royal ban on coal, "great fines and ransoms": Galloway, *History,* 10.

Page 3: "Monsters of the vegetable world": Nicolls, *Story of American Coals,* 32-33.

Page 5: Humans most efficient at turning calories into energy: Debeir, 4.

Page 5: Fire's impact on humanity generally: Goudsblom. Fire's use as a distinguishing human feature: Goudsblom, 17. Fire's role in human evolution and social development: Patterson, 52, 61. Fire controlled half million years ago: Debeir, 15.

Pages 5-6: Clearing land by fire, and fire's importance to agriculture and permanent settlements: Goudsblom, 28, 44-54, 58-65.

Page 6: Charcoal formation: Richard H. Schallenberg, "Charcoal Iron: The Coal Mines of the Forest," in Hindle, 290, 294.

Page 6: Limits on the capture of solar energy limited economies: Wrigley, 51.

Page 10: "Every basket is power and civilization": Emerson, 86-87.

Page 10: "With Coal we have light": Nicolls, *Coal Catechism,* 6-7.

Pages 10-11: Coal short story in Charles Dickens's journal, and "man may hereafter live": "The True Story of a Coal Fire," in *Household Words* 1 (1850): 26-90 (in three parts); quote at 29-30.

Page 12: "A race of men": "The History and Destiny of Coal," in *Christian Review* 21 (April 1856): 267-283; quote at 282.

Page 12: Missionary focused on coal: Williamson, passim. "Cause China's rising sun" and "bound and black and mighty": Gibson, 6, 59.

Chapter Two: The Best Stone in Britain

Page 15: "Best stone in Britain," Roman use of jet: Galloway, *Annals,* 1:6-7.

Pages 15-16: Burned by Romans in Britain: Nef, 1:2. Use at temple of Minerva: Hatcher, 17. Use by Welsh for cremations: Hatcher, 17.

Pages 16–17: St. Bede: Galloway, *History,* 2–3.

Page 17: Plants ashore 425 million years ago: Cleal and Thomas, 1. Spread of plants to land and rise of Carboniferous forests: Reader, 67–85.

Pages 17–18: A reconstruction of the continents' presumed locations during this time can be seen at the U.S. Geological Survey Web site: http://wrgis.wr.usgs.gov/docs/usgsnps /pltec/sc306ma.html. Lepidodendron described, likely 175 feet (54 meters) tall: Stewart, 103–104. Lepidodendron's yard-long leaves: Taylor and Taylor, 248.

Page 18: Lepidodendron possibly covered with leaves, and with pithy interiors: Ralph E. Taggart, "Carboniferous Forests," Michigan State University Department of Botany and Plant Pathology, article available at http://taggart.glg.msu.edu/isb200/carbfor.htm.

Pages 18–19: Sigillaria described: Stewart, 119. "Like a huge barrel": Knowlton, 89. Ancient giant horsetails: Taylor and Taylor, 320. Giant ferns: Cleal and Thomas, 52–53.

Pages 19–20: Foot-long cockroaches: Taggart, *supra.* Giant dragonflies and millipedes: Maurice E. Tucker, "Life in the Swamp: The Fauna and Flora of the Carboniferous Coal-Forming Environments," Presidential Address for the British Association for the Advancement of Science, Sept. 1995, Newcastle. "As long as a cow": Attenborough, 62. Amphibians, belly-prints: Tucker, *supra.* Evolution from amphibians to reptiles during the Carboniferous: Reader, 90–92.

Page 20: Burial process: Carpenter and Astwood, 123. Sea levels rising and falling, swamping the forests: Tucker, *supra.*

Page 20: Coal formation: Carpenter and Astwood, 123.

Page 21: Origin of term *sea-coal*: Nef, 2:452.

Page 21: Exposed coal along Tyne: Hatcher, 70.

Page 22: Shipping costs, and Tyne flowing "with solemn majesty": Levine and Wrightson, 9, 11.

Page 22: Coal lands owned by church: Nef, 1:134–135.

Page 23: Legal status of coal: Nef, 1:281.

Page 24: Skirmish with monks, and larger struggle between merchants and church: Galloway, *History,* 8; Hatcher, 521.

Page 24: Queen Eleanor: Galloway, *Annals,* 1:26–27; Brimblecombe, 9.

Page 25: Coal was particularly smoky: Nef, 1:12–13. Lack of chimneys, interiors filling with smoke: Nef, 1:199; Hatcher, 40; Holland, 319. Fires in stick huts of Homo erectus: Patterson, 56.

Page 25: "Infected and corrupted": Hatcher, 25. General anticoal revolt in 1306: Galloway, *Annals,* 1:30. Royal commissions and proclamation: William Te Brake, "Air Pollution and Fuel Crisis in Preindustrial London, 1250–1650," *Technology and Culture,* vol. 16, no. 3 (July 1975): 339–340. Execution of a ban violator unlikely: Brimblecombe, 9.

Page 26: Wood shortage: Hatcher, 19–20. Industries turning to coal: Brimblecombe, 17.

Page 26: The spread of the Black Death, population weakened by unstable climate: Gribbin and Gribbin, 142; Tuchman, 24, 92–125. University of Paris theory: Tuchman, 103.

Page 27: Half as many people: Hatcher, 28. Forests rebound postplague: Te Brake, *supra*.

Page 27: Buboes were like "fragments of brittle sea-coal" and "like a burning cinder": Tuchman, 93. "Anthrax" Greek for coals: Nef, 1:2, n. 6.

Page 27: "Death coming into our midst": Tuchman, 93.

Page 28: Coal industry limited under church ownership: Nef, 1:135.

Page 29: Fifth of the land, three times the Crown's income: Smith, 13.

Page 29: Three-century battle between merchants and church: Nef, 1:140–156.

Pages 29–30: England relative to Europe in mid-1500s: Trevelyan, 134–135. Semi-colonial relations, not urban: Sharpe, 141 and 84. Small merchant marine and navy: Nef, 1:238.

Page 30: "Good round log": Trevelyan, 129–130. Sheep pastures displacing forests: Nef, 1:190. Iron industries using up forests: Galloway, *Annals,* 1:79.

Pages 30–31: Elizabethan commissions, "greatly decayed and spoiled": Nef, 1:158–159. Wood shortage, generally: Hatcher, chapter 3. Hedgestealers "whipped till they bleed well": Hatcher, 49.

Page 31: Rising wood prices, 20,000 wagon loads to breweries: Nef, 1:158, 192–193. Price of wood: Hatcher, 37.

Pages 31–32: Little Ice Age: Gribbin and Gribbin, 141, 151; Tuchman, 24. Frozen Thames: Gribbin and Gribbin, 150.

Page 32: London deaths outpacing births, later marriages during difficult economic times: Sharpe, 85, 40.

Pages 32–33: Surge in domestic coal use, "nice dames of London," Galloway, *History,* 22–25.

Page 33: London's population, London's and England's rising status: Sharpe, 85, 87.

Pages 33–34: Chimneys common by mid-1500s, and belief that wood smoke protected health: Galloway, *History,* 23. Children chimney sweeps: Hammond and Hammond, 2:12–19.

Page 34: Elizabeth "greatly grieved and annoyed": Brimblecombe, 56. Hugh Platt's book: Nef, 1:247; Brimblecombe, 31.

Page 35: "The City of London Resembles the . . . Suburbs of Hell": Evelyn, 19. "That of all the Cities perhaps of Europe": Nourse, 349.

Pages 35–36: Descriptions of smoke, "sooty Crust," more damage in one year: Evelyn, 19–20. Buildings "peel'd and fley'd": Nourse, 350.

Page 36: "Insinuating itself into our very secret Cabinets" and "black and smutty Atomes": Evelyn, 20. Tapestries "stinking richly": Nourse, 351.

Page 36: "In a word, 'tis impossible for any Man to live sweet and clean": Nourse, 364.

Page 37: Londoners begin to carry umbrellas: Brimblecombe, 63–64. "When Men think to take the sweet Air": Nourse, 350. Soot in the Thames: Evelyn, 32–33.

Page 37: "Those few wretched fruits": Evelyn, 21. "So that everybody in London": Brimblecombe, 68.

Pages 37–38: "Coughing and Snuffing" and "the Barking and the Spitting," voices changing, visitors getting sick, lung ailments "rage more in this one City": Evelyn, 26, 24, 18.

Page 38: Soot "killing multitudes": Evelyn, 30. "From this stinking and smoaky Air": Nourse, 364.

Page 38: Miasmatic theory: Brimblecombe, 9; Howe, 186–187.

Page 39: Smoke and plague mortality: Evelyn, 25. Coal fires recommended to help purify air: "Necessary Directions for the Prevention of the Plague in 1665, With Divers Remedies of Small Charge, by the College of Physicians," in Anonymous, *A Collection of Very Valuable and Scarce Pieces Relating to the Last Plague in 1665,* 36–40.

Page 40: "Ancient matrons," data collection: Graunt, 17–19, 26. "If they found any more such tradesmen": Brimblecombe, 52.

Page 40: Death rates and causes: Graunt, 29–30.

Pages 40–41: Coal thought to cause higher death rates, residents live long "yet new-comers, and Children," and smoke holds back miasmatic air: Graunt, 76, 56, 74.

Pages 41–42: Fuel as tenth of income, fire for part of day: Nef, 2:203–204.

Page 42: Repeated fears that people were on brink of violent revolt: Nef, 2:206. Gardens thrived during blockade: Evelyn, iv. "Great and unspeakable" complaints: Galloway, *History,* 35. Many poor perished from lack of fuel: Nef, 2:204.

Chapter Three: Launching a Revolution

Page 44: "Whereas when we are in London": Daniel Defoe, quoted in Levine and Wrightson, 80. Farming economy comes to end in Newcastle: Levine and Wrightson, 83.

Page 45: Star Chamber proceeding, "lewd persons" and other quotes: Nef, 2:150–151.

Page 45: "New gulf between the classes": Nef, 2:166. Miners' different habits and speech: Ashton and Sykes, 70.

Page 46: Scottish miners in bondage: Ashton and Sykes, 71–74.

Page 46: "Torture in the irons": Hatcher, 312. The original quote used the term *colliers* instead of *coal miners.*

Page 46: Higher wages for miners: Ashton and Sykes, 77.

Page 47: Conclusion that coal mining was punishment for original sin: Holland, 288, quoting the Rev. T. Gisborne.

Pages 47–48: Choke damp described: Galloway, *Annals* 1:160; Hatcher, 233.

Page 48: "Fell down dead," "God's mercy," and "lightness of Brain" quotes, falling off ropes: Nef, 2:170–171. Description of the "ordinary remedy": Galloway, *History,* 70.

Pages 48–49: "An odor of the most fragrant kind": Nicolls, *The Story of American Coals,* 173.

Page 49: Pink-snouted, crouching mice less useful than toppling canaries: Pohs, 261. Customary to lower a dog: Galloway, *History,* 69. Dog lowered only after first human death: Nef, 2:173.

Page 50: "With the noise of loudest thunder": quoted in Flinn, 129. Victims shot from mines like bullets from a gun: Nef, 1:363; Galloway, *History,* 88.

Pages 50–51: Description of fireman duties: Galloway, *History,* 72. "The miner then draws near to the fire, and frightens it with his staff": Eavanson, 28–29, quoting Nicander Nucius, 1554.

Page 51: Fish in a phosphorescent state: Galloway, *History,* 124.

Pages 51–52: "As we have been requested to take no particular notice": Galloway, *History,* 107–108.

Page 52: 1833 river floods into Scottish coal mine, "a slight eruption," and "burst through the surface": Holland, 252.

Page 53: 1610 prediction: Galloway, *Annals,* 1:128.

Pages 53–54: Thousands of drainage tunnels by mid-1600s: Nef, 1:354. Some only eighteen inches wide and four feet high: Nef, 2:449.

Page 54: Tunnel systems described, one five miles long: Flinn, 110–111. Risk of being "dashed in pieces": Nef, 2:170, quoting from 1665–1666 Royal Society Phil. Trans.

Pages 54–55: Laborers bailing out the mine with buckets: Nef, 2:449. Windlasses discussed: Hatcher, 217–220. Chain of buckets described, "fell to the bottom with a most tremendous crash": Galloway, *History,* 77. Description of vacuum pump: Hatcher, 223–224.

Page 55: Different pumps working together: Nef, 2:451. Large mines needed fifty to sixty horses: Hatcher, 227.

Page 56: Possibly ten-fold growth between 1550 and 1700: Hatcher, 47. More energy than possible from all England's woodlands: Hatcher, 55. Five times more coal than rest of world: Wrigley, 54.

Page 56: "Unwrought or drowned": Hatcher, 231, quoting the *Compleat Collier,* 1708.

Pages 56–57: Royal Society description, members, "religion, nationality, profession": Nef, 1:252–253.

Page 57: "Special seeds": Wendt, 346. Biblical flood theory, "whole Terrestrial Globe," and end of May determination: Wendt, 27; Woodward, *An Essay Toward the Natural History of the Earth,* preface; Woodward, *An Attempt Towards a Natural History,* 108–9. Boyle quote: Nef, 2:175, text and n. 3.

Page 58: "New Digester or Engine for softening Bones" and cooking supper: Rolt and Allen, 24; Smiles, *Lives of Boulton and Watt,* 32.

Page 58: Papin's demonstration, abandonment of steam: Rolt and Allen, 24.

Pages 58–59: Royal Society snubbed Newcomen, "to the small scientific world": Rolt and Allen, 12.

Page 59: Evidence that Newcomen had idea himself: Rolt and Allen, 38–39.

Page 59: Description of engine: Rolt and Allen, 41.

Pages 59–60: First engine at coal mine: Briggs, *The Power of Steam,* 32. Fifty horses replaced: Galloway, *History,* 82.

Page 60: "The most wonderful invention": Galloway, *History,* 80. Giving "to the whole world": Rolt and Allen, 11.

Page 60, footnote: Newcomen's grave lost: Rolt and Allen, 11.

Page 60: Hundreds of Newcomen engines in use: Briggs, *Power of Steam,* 50–51. Dimensions of early engines: Rolt and Allen, 108.

Page 61: Watt's childhood, instrument-making: Smiles, *Lives of Boulton and Watt,* 86–92.

Pages 61–62: Repairing model, walk in the park: Briggs, *The Power of Steam,* 50–52. Joseph Black's "this capital improvement" quote: Smiles, *Lives of Boulton and Watt,* 129.

Page 62, footnote: Black's discovery of carbonic acid gas: Smiles, *Lives of Boulton and Watt,* 372.

Page 62: Unsuccessful early efforts, "nothing more foolish than inventing": Smiles, *Lives of Boulton and Watt,* 184–189, 150.

Page 63: Lunar Society: Schofield.

Page 63: Soho described: Schofield, 26–28 and Smiles, *Lives of Boulton and Watt,* 169–176.

Pages 63–64: Water problems at Soho: Smiles, *Lives of Boulton and Watt,* 182. Correspondence with Franklin: Schofield, 60. Transfer to Boulton: Smiles, *Lives of Boulton and Watt,* 197. Wilkinson manufactures cylinder: Smiles, *Lives of Boulton and Watt,* 212. Two 1776 engines: Briggs, *The Power of Steam,* 54.

Page 64: Royalties related to fuel savings, Watt becomes national hero: Briggs, *The Power of Steam*, 55, 50.

Page 64: "I am engaged, your Majesty": Smiles, *Lives of Boulton and Watt*, 3–4.

Page 65: Smelters depleting the woods: Nef, 1:193–195. British importing steel: Thomas, 13, 74.

Pages 65–66: Adoption of coal in iron smelting and refining: Flinn, 240–241.

Page 66: Coke and steam-powered blast allowed bigger furnaces: Thomas, 103. Britain's iron industry most efficient in world: Landes, 95.

Pages 66–67: Reinforcing relationship between iron, steam and coal: Briggs, *The Power of Steam*, 35, 39.

Page 67: Coal's expansion, 1700–1830: Flinn, 442. Industry growth after 1830: Tuttle, 141.

Page 67: Ancient use of waterwheels: Smil, *Energy in World History*, 103. Persistence of word "mill": Mumford, 138.

Page 68: Steam and coal concentrate production: Mumford, 161–162; Wrigley, 75–76.

Page 68: Industrial revolution like fire, Tree of Knowledge, wheel: Landes, 5, 12, 42. 1780 to 1830 called the conventional chronology of industrial revolution: Wrigley, 9.

Page 69: British monopoly over industrial revolution scarcely challenged for at least half a century: Stearns, 41. Britain produces four-fifths of world's coal: Stearns, 30. Produces more iron than rest of world: Landes, 95.

Chapter 4: Full Steam Ahead

Page 72: "From this foul drain": Tocqueville, 107–108; quoted in Marcus, 66.

Page 72: "Nature can be conquered": Briggs, *The Power of Steam*, 72.

Page 72: "Nearly 500 chimneys": Wohl, 208–209.

Page 74: Big cotton mills in Manchester: Rule, 10.

Page 75: "Whilst the engine runs": Ward, 26–27.

Page 76: Workers had formerly led lives without clocks: Hammond and Hammond, 1:33.

Page 76: Gas lighting in 1792: Robertson, 34. Gas lights introduced in cotton mills in 1805: Smil, *Energy in World History*, 160. Palace comparison: Marcus, 39.

Page 76: Gas lights may have actually shortened days: Marcus, n. 34.

Pages 77–78: Exploitation of children in the mines, "in all the coal mines," "I have to trap without a light," and "chained, belted, harnessed like dogs": Hammond and Hammond, 1:39, and 2:8, quoting the *First Report of the Commission for Inquiring Into Employment of Children in Mines and Manufactures, 1842*.

Page 78: Comparing child labor in cotton mills with mines: Tuttle, 96, 142.

Page 78: Young people with income leave home: Gaskell, 98–99. Rise in population due to earlier marriages: Hudson, 135.

Page 79: "A fresh race of beings": Thompson, 190. New class "but a Hercules in the cradle": Gaskell, 6. Steam engine "entered on no prepared heritage": Cooke Taylor, 4.

Page 79: "The conversion of a great people": Gaskell, 10. Majority urban: Rule, 16.

Pages 79–80: Manchester becomes a city in 1853: Briggs, *Victorian Cities,* 108. Trade unions banned: Jevons, 445–488; Rule, 266.

Page 80: Engels's father's business, Engels's expectation of revolution, his double life in Manchester: Marcus, 67, 88, 91–92. Engels's book attracted considerable attention: Engels, introduction, by David McLellan, xvi. Relationship with Marx: Marcus, 113, 121.

Page 81: "It is scarcely in the power of the factory workman": Hammond and Hammond, 1:56.

Page 81: Smoke "risen to an intolerable pitch": Wohl, 208–209. Smoke forms "an inky canopy": Cooke Taylor, 2.

Pages 81–82: "It is an appalling fact" and comparison of Manchester life expectancies with rural life: Chadwick, 223.

Page 82: "Robust and well-made" men and "vast deterioration": Gaskell, 158, 160–162. Queen Victoria's observations, "nobody moved," and "painfully unhealthy-looking": Briggs, *Victorian Cities,* 109. Recruits rejected for physical weakness: Rule, 381.

Page 83: Description of the symptoms, causes and consequences of rickets: Ferguson; Michael F. Holick, "Photosynthesis, Metabolism, and Biologic Actions of Vitamin D," in Glorieux.

Pages 83–84: "A disc without rays": Tocqueville, 107. Every child in some industrial neighborhoods, "the English disease": Wohl, 56, 57. Rickets affected half the industrial population, was "probably the most potent factor": Ferguson, 9, 18.

Page 84: Adequacy of coal-gas lights: Ure, 374–375; discussed in Mumford, 169.

Page 85: "It is the great quantities": Hatcher, 459. Growth of the shipping fleet, more ships moving coal than anything else: Nef, 1:238. "The coal trade may be regarded": Nef, 1:240.

Page 86: Armada defeated with aid of merchant ships: Davis, 45. Coal trade was "chief nursery" of seamen: Nef, 1:238, Galloway, *History,* 34; Hatcher, 472. Coal vessels more useful than fishing vessels: Nef, 1:239. Coal ships and crews pressed into service: Flinn, 173; Hatcher, 478–479.

Pages 86–87: Coastal coal trade enjoyed almost "superstitious reverence": Nef, 2:206. Opposition to mining closer to London: Galloway, *History,* 35.

Page 87: Royalty traveled on horseback: Ransom, 6. Coal carts "struck . . . terror into the poor country people": Flinn, 147. Birmingham roads: Smiles, *Lives of Boulton and Watt*, 162. Trip from Glasgow to London: Dickerman, 9.

Pages 87–88: Canal construction to carry coal, duke of Bridgewater's canal, "must have coals": Nef, 1:258; Flinn, 181–183; Ransom, 10. Coal one reason for turnpikes: Flinn, 147–148.

Page 88: Horse-drawn railways: Ransom, 35.

Pages 88–89: History of rails: Ransom, 12–13. Newcastle area wagonways: Levine and Wrightson, 49–76; Flinn, 149. Boy who would "free the steam engine": Rolt, 5.

Page 89: Stephenson's childhood, early work: Rolt, 6–12.

Pages 89–90: Learning to read: Smiles, *Life of George Stephenson*, 32. Personal misfortunes: Rolt, 6–12.

Page 90: Relationship with Grand Allies, focus on moving coal: Rolt, 4, 20; Smiles, *Life of George Stephenson*, 84.

Pages 90–91: All the elements of a railway: Ransom, 28.

Page 91: Stockton and Darlington railway: Ransom, 39–40.

Page 91: Liverpool and Manchester first to move solely by locomotive: Ellis, 18. Contest settled question of horses, cables or locomotives: Ransom, 45–53.

Page 92: "A huge monster in mortal agony": Freeman, 38. The clerk "fell prostrate": Simmons, 15–16.

Pages 92–93: "Swifter than a bird flies," "I stood up," and "most horribly in love": Kemble, 162–163. Kemble's also ride described in Rolt, 190–192.

Pages 93–95: This account of the opening day of the Liverpool and Manchester Railway is mainly compiled from Ellis, 17–19; Ransom, 55–56; Freeman, 30–31.

Page 94: "Completely lost his head": Kemble, 2:190.

Page 95: Wellington's attitudes toward "lower orders": Wellington, 95, letter to Mrs. Arbuthnot, 1 May 1831. Later made money in railroad stocks: Wellington biography available at Spartacus Web site: http://www.spartacus.schoolnet.co.uk/PRwellington.htm.

Page 95: Ridership of 700,000 in months ahead: Ellis, 31. Miles of track laid: Freeman, 1. Rails as harbinger of brighter future, and "a great, a lasting, an almost perennial conquest": Williams, 285–286, 27.

Page 96: Manchester and London population statistics: Hudson, 152; Briggs, *Victorian Cities*, 324, 331. London "the center of the . . . world," 1883 quote from U.S. writer: Briggs, *Victorian Cities*, 328.

Page 96: London "a volcano with a hundred mouths": Brimblecombe, 113.

Pages 96–97: English abhorred stoves, loss of visible fire, indoor air quality concerns: "Coal and Smoke," in *Littells Living Age* 89 (May 1866): 26. Most homes still warmed by open fires in 1920, energy wasted: Robertson, 23.

Page 97: "For the greater part of the day": *Times* (London), 11 January 1812, p. 3.

Page 98: "Great Stinking Fogs" recorded in late 1600s, increase in fogs, role of pollution, nicknamed "the Big Smoke": Brimblecombe, 109–113. Byron's "wilderness of steeples": quoted in Brimblecombe, 91.

Page 98: Smoke as "sublime canopy": Briggs, *Victorian Cities,* 321. "Beloved smoke": Brimblecombe, 85.

Pages 98–99: Smoke "gave a kind of solidity and nutriment": Brimblecombe, 85–86. "Today we are having a yellow fog," and variously colored fogs: Brimblecombe, 117, 125.

Page 99: "All locomotion . . . extremely dangerous": *Times* (London), 10 December 1873, p. 7.

Pages 99–100: Parade of fog-related accidents: *Times* (London), 10–13 December 1873. Effect on cattle: Brimblecombe, 123.

Page 100: Two deaths from "inhaling the fog": *Times* (London), 12 December 1873, p. 4; 13 December 1873, p. 7. Deaths during fogs: Brimblecombe, 124 (chart).

Chapter 5: A Precious Seed

Page 103: Forests not seen in their homeland for thousands of years: In 4000 B.C., most of Britain was covered with forests except for a few areas. Neolithic man deforested the land for agriculture; by 500 B.C., only half of England remained wooded: Morin, 23. Jamestown history, sending home cedar instead of gold, and "wooded to the brink of the sea": Morgan, 116, 139.

Pages 103–104: First Puritan reverend: Morgan, 166–167. "A poor servant here": "Francis Higginson's New-England's Plantation," in Young, 254.

Page 104: "Extraordinary clear and dry air": "Francis Higginson's New-England's Plantation," in Young, 251. Death of Higginson and other members of colony: Morgan, 167, 168.

Page 104: Scale of eastern American forest: Morin, 53; Perlin, 9, 249–274, 324–326.

Pages 104–105: "A hideous and desolate wilderness" and "snakes and serpents": Morgan, 134. Settlers' fear of wilderness, generally: "Francis Higginson's New England's Plantations," in Young, 255; Perlin, 270. Native influence on the forest: Morin, 54.

Page 105: Early wood consumption: Perlin, 349–352. Coal field half the size of Europe: Holland, 472.

Pages 105–106: "Scattered by the hand of the Creator with very judicious care, as precious seed": "The History and Destiny of Coal," *Christian Review* 21 (April 1856): 267.

Page 106: "Coal is also in abundance": Baldwin, 6. "I was truly astonished": Fleming, 564, quoting Lyell following 1846 visit to the Ohio valley.

Pages 106–107: Appalachian field reaches 190 miles at Pittsburgh, stretches to Alabama: Shurick, 17. Forks "much infested with venomous Serpents and Muskeetose": Baldwin, 5.

Page 107: Brutal battles at the Forks: Henry Steele Commager, "Forts in the Wilderness," in Lorant, 22–26; Baldwin, 32–52. "This valuable Acquisition": Commager, in Lorant, 34, quoting anonymous correspondent to the *Pennsylvania Gazette.*

Page 107: Early mining in Pittsburgh: Commager, in Lorant, 34; Baldwin, 62; Eavenson, 51.

Page 108: Pittsburgh's future and "vegetable air": Stefan Lorant, "Gateway to the West," in Lorant, 52. 1790 pollution "by reason of using so much coal" and population: Lorant, in Lorant, 55.

Page 108: Pittsburgh's early industries: Baldwin, 145–149. Early industries and population, "Birmingham of America": Lorant, in Lorant, 69, 77, 79.

Pages 108–109: "Dark, dense smoke": Lorant, in Lorant, 78. "A cloud which almost amounts to night," "an immense column of dusky smoke," and "stained, soiled and tarnished": Jakle, 135–136.

Page 109: "Even the filth and wondrous blackness": Trollope, quoted in Baldwin, 202.

Page 110, footnote: Lack of steam power outside of Pittsburgh: Alfred D. Chandler, Jr., "Anthracite Coal and the Beginnings of the Industrial Revolution in the United States," reprinted in Church, 398–399.

Page 110: Pittsburgh as steam capital of the continent, eastern cities getting coal from other sources, industrial revolution stifled in East: Chandler, in Church, 404–405.

Page 111: "the wild place": Miller and Sharpless, 2. "Switzerland of America": Korson, 20. "The mountain was on fire": Miller and Sharpless, 9.

Page 111: Pennsylvania coal formation, oceans rising and falling, leaving layers of coal: McPhee, 246.

Pages 111–112: Rise of the Appalachians: McPhee, 126; Sullivan, 279. Appalachians on the scale of the Himalayas today: Sullivan, 267. A reconstruction of the continents' presumed locations during the Carboniferous is available at the U.S. Geological Survey Web site: http://wrgis.wr.usgs.gov/docs/usgsnps/pltec/sc306ma.html.

Page 112: Anthracite region took the brunt of the collision: McPhee, 247. Much of world's anthracite in five Pennsylvania counties: Miller and Sharpless, 5.

Page 112: Purchase of anthracite region: Nicolls, *The Story of American Coals,* 71. Use to make paint: Korson, 2.

Pages 112–113: History of Summit Hill, depth of outcrop: Korson, 1–31, 50. Would be considered most magnificent coal property: Richardson, 47.

Pages 113–114: Account of 1803 arks, "now the torrent roars," "to the curious speculation," used to gravel footwalks: Nicolls, *The Story of American Coals,* 60–61.

Pages 114–115: "If the world should take fire" and 1859 Summit Mine fire: Korson, 34, 19.

Page 115: Jacob Cist's efforts with "good Latin scholar" etc., "the whole thing was a romance," Korson, 38–39.

Pages 115–116: Three arks sink, fourth arrives with crew "glad to be alive," coal sold to White and Hazard: Korson, 36, 42.

Page 116: "Gentlemen, you have our permission": Korson, 44.

Page 116: Josiah White in buckskins: Richardson, 55. Brings a thousand laborers: Korson, 45.

Pages 116–117: Cobbing and whiskey rationing: Korson, 45. Drunken Sundays, partner armed: Miller and Sharpless, 23–24. "No inducements upon them": Richardson, 55.

Page 117: Laboring in the cold water: Miller and Sharpless, 24. Twelve small dams, cut journey to day and a half: Korson, 52–53.

Page 117: Price of firewood high in Philadelphia: Morin, 55.

Page 118: Baldness, tooth decay, etc., promotion campaign: Korson, 53, 57–58.

Page 118: Built arks in forty-five minutes, running out of trees: Miller and Sharpless, 25. Sold wood, walked back: Korson, 411, footnote 40.

Page 118: "from this port . . . there is a fleet of 400 vessels": Korson, 109.

Page 119: Walk to Philadelphia, ruined boots: Waggoner, 192. Men pulling coal boats, canal ahead of the Erie: Korson, 111.

Pages 119–120: Other canals: Korson, 109–110. Lehigh and Delaware and Hudson canals: Miller and Sharpless 26, 36–37. Delaware and Hudson subscription and bailout: Delaware and Hudson Company, *A Century of Progress,* 23, 35.

Page 120: Canal workers, tools, conditions, formed largest canal network in nation: Miller and Sharpless, 26–27, 38. Busiest waterway in the United States: Miller and Sharpless, 34; Korson, 109.

Pages 120–121: Schuylkill canal company couldn't mine coal: Bogen, 11. Coal rush of 1829: Korson, 98; Miller and Sharpless, 46–47, 87. Production rose for a century: Shurick, appendix, table 8, and 357.

Pages 121–122: Cradle of railroading; first to build lines: Korson, 122. Summit mine mule-driven line: Carter, 36.

Page 122: Schuylkill County more rail than anywhere, rail and coal considered one industry: Bogen, 11–15.

Page 122: Use of wood by early railroads: John H. White, Jr., "Railroads: Wood to Burn," in Hindle, 199–201.

Pages 122–123: Brightly painted engines, well-dressed engineers: White, in Hindle, 199, 203.

Page 123: "a storm of fiery snow": White, in Hindle, 201.

Page 123: Mohawk Valley trip, "a general melee": Adams, 49.

Pages 123–124: Money burned, women "almost denuded": White, in Hindle, 201. Buckets of sand: Waggoner, 198.

Page 124: Charcoal-iron production in the United States: Lorant, 145; Perlin, 337–338. Factories held back by lack of cheap iron: Chandler, in Church, 402. Railroads held back: Miller and Sharpless, 59.

Page 125: Thomas and furnace on the Lehigh: Miller and Sharpless, 60.

Pages 125–126: Rise of mass production between 1835 and 1855: Chandler, in Church, 430.

Page 126: Change in locus of industry, rise of worker and manager, increase of manufacturing from 17 percent to 30 percent in 1840s: Chandler, in Church, 429–432.

Pages 126–127: Agrarian south does not industrialize: Hindle and Lubar, 166. Industrial ratios, north and south: P. Johnson, 462.

Page 127: Railroads to ship surplus made western agriculture and ranching profitable: P. Johnson, 514–517.

Pages 127–128: Incompatibility with Jeffersonian ideal, "manufacturing breeds lords": Hindle and Lubar, 57, 92.

Chapter 6: The Rise and Fall of King Coal

Page 129: Opening day of the centennial: William H. Rideing, "At the Exhibition, Part II," in *Appleton's Journal* 15, no. 370 (1876): 759; D. Brown, 128–129.

Page 130: Molly Maguires in Ireland: Miller and Sharpless, 138–139; Wallace, 323.

Page 131: "well-nigh irresistible": Bogen, 52.

Pages 131–132: Gowen's philosophy of growth, gospel of "bigness": Bogen, 52; Schlegel, 61. Gowen sees union as threat, raises freight rates to fight it: Schlegel, 17–20. Sneaks law through legislature: Schlegel, 34. Buys up mines: Miller and Sharpless, 154.

Page 132: Organizes "pool," reported in the newspaper: Schlegel, 44.

Pages 132–133: Long strike, violence on both sides, miners blamed Gowen: Miller and Sharpless, 156–158; Schlegel, 63–76. "With one hand reaches for the pockets of the consumers": Schlegel, 68.

Page 133: Legislative scrutiny: Aurand, 99. Gowen shifts attention to "class of agitators": Schlegel, 84–86.

Page 133: Pinkerton and his agent: Schlegel, 87–99. Pinkerton, Coal and Iron police, 1875 roundup of suspects: Miller and Sharpless, 155–159.

Page 134: Trial, Gowen as special prosecutor: Miller and Sharpless, 165–166.

Page 134: Press links Mollies to union, numbers hanged and imprisoned, and "one of the greatest works": Schlegel, 126, 149, 151.

Pages 134–135: Mollies in national folklore: Schlegel, 150–154; Miller and Sharpless, 137. Union prevented crime, labor linked to terrorism in public opinion: Aurand, 99, 108–109.

Page 135: Stocks and bonds held in England: Bogen, 55. Provokes railroad war, alliance with Vanderbilt: Schlegel, 179, 226.

Pages 135–136: Morgan forces Gowen out: Schegel, 254–269.

Page 136: "Papa is simply triumphant": Strouse, 252. Gowen's suicide, obscurity: Schlegel, 286–288.

Page 137, text and footnote: United States, Britain, German coal and iron production: Degler, 29. Coal production statistics: Schurr, 69.

Pages 137–138, text and footnote: Bituminous and anthracite production: Schurr, 63. U.S. coal fields: Shurick, 16–27. Thousands of bituminous mines in thirty-three states: J. Johnson, 13; Seltzer, 35.

Page 138: 1898 agreement: Craig Phelan, "John Mitchell and the Politics of the Trade Agreement, 1898–1917," in Laslett, 78. UMW for forty years the biggest, most powerful union: John H. M. Laslett, "A Model of Industrial Solidarity? Interpreting the UMWA's First Hundred Years, 1890–1990," in Laslett, 1.

Page 138: Mitchell in mines at twelve: Phelan, 6–7. Rise within UMW: Phelan, in Laslett, 75.

Pages 138–139: Conditions worse in anthracite country: Gluck, 67–68. Breaker boys: Greene, 217–220. Mules: Miller and Sharpless, 102.

Page 139: "It is a common site," and sharing lunches with rats: Long, 35–36, Miller and Sharpless, 130. Bigger, meaner, uglier rats: Greene, 212.

Pages 139–140: Immigrants lured by companies: Miller and Sharpless, 172–173. UMW efforts to enlist Europeans: Phelan, in Laslett, 83. McKinley assassination and weeping crowds: Phelan, 178.

Page 140: 1902 strike, "the best-managed": Phelan, in Laslett, 88. Public sympathy, "the prince of moderation": Phelan, 359.

Page 141: George "Divine Right" Baer: Phelan, 180. Roosevelt on Baer: Levy, 31.

Page 141: Pressure on Morgan, settlement hailed as victory: Phelan, 185–189. First time president intervened for workers: Painter, 186. Mitchell's legacy, "new model for union leadership": Phelan, in Laslett, 100–103.

Pages 141–142: Arcola coal riot: *New York Times,* 11 January 1903, p. 1.

Page 142: "roaring, hilarious voice of invitation" and "barefoot and bleeding": Strasser, 53.

Pages 142–143: Stoves common after Civil War: Strasser, 36.

Page 143: Sear's stoves: Described by Strasser, 38–39. Asbestos sheets used as insulation: Van Rensselaer, 211.

Pages 143–144: Lighting coal stove: Kinne and Cooley, 46; Holt, 282; Van Rensselaer, 210.

Pages 144–145: Boston experiment: Strasser, 41. Graphite formation: Bruere, 6; Nicolls, *Story of American Coals,* 56.

Page 145: Coal stove too hot four months of year, coal had to burn on once lit: Holt, 14. Recommended kerosene stove in summer: Van Rensselaer, 207.

Pages 145–146: Tons of coal needed per winter: Beecher and Beecher-Stowe, 361. "Not only poison their families": Beecher and Beecher-Stowe, 81.

Page 146: Concerns over vitiated air: Beecher and Beecher-Stowe, 81, 421; Stradling, 47–48.

Page 147: Fatal consequences if gas-light keys leaked: Holt 270. "Gasoliers": Robertson, 34.

Page 147: Rise of kerosene and Standard Oil: Yergin, 50–51. Kerosene had largely replaced sperm whale oil by 1869: Beecher and Beecher-Stowe, 363.

Page 148: Battles for market between coal and oil: Yergin, 543.

Page 148, text and footnote: Bituminous cities, oil in West and Southwest: Stradling, 15, 183. Natural gas in San Francisco: R. Dale Grinder, "The Battle for Clean Air: The Smoke Problem in Post–Civil War America," in Melosi, 84. Coal prices in St. Louis: Stradling, 9.

Pages 148–149: People using bituminous during strike: *New York Times,* 4 June 1902. "Smoke Pall Hangs Over the Metropolis": *New York Times,* 7 June 1902. "Are we to have fastened on us": *New York Times,* 15 June 1902. "If New York allows bituminous coal to get a foothold": Stradling, 17.

Pages 149–150: Women's clubs, "municipal housekeeping": Stradling, 42–45; Angela Gugliotta, "Class, Gender, and Coal Smoke: Gender Ideology and Environmental Justice in Pittsburgh, 1868–1914," in *Environmental History* 5, no. 2 (April 2000): 165–193.

Page 150, footnote: "Gaseous sewage" idea: "Coal and Smoke," in *Littell's Living Age* 89 (May 26, 1866): 515, 528–529.

Pages 150–151: "given women's roles as keepers of the house": Stradling, 49. Accusations of frivolity, desire to protect complexions: Gugliotta, *supra,* 174.

Page 151, text and footnote: Alliances with other groups, unsentimental approach, seventy-five cities with smoke abatement ordinances: Stradling, 52–55 and Gugliotta, *supra.* 1909 cartoon: Stradling, 52–53.

Pages 151–152: Focus on germs made smoke's link to disease weaker: Gugliotta, *supra,* 177. Birmingham physician's belief that smoke was purified by fire: Stradling, 135. Chicago coal dealer's belief that soot filtered air: Stradling, 208, n. 14.

Page 152: Acne, diarrhea, constipation, and "pale and flabby": Stradling 45, 50. "Chicago's black pall of smoke": Grinding, in Melosi, 86.

Pages 152–153: Gloom causes women "to be irritable," husbands to drink, makes children "dull, apathetic": Grinding, in Melosi, 86. Pittsburgh psychologist, "chronic ennui," lack of "clear, trenchant, reflective thinking": Stradling, 32–33.

Page 153: Cincinnati death rates: Stradling, 25. German and English data and Pittsburgh findings: Stradling, 50–51, 108–109. "Our most useful business men": Gugliotta, *supra,* 181.

Pages 153–154: Smoke's economic costs: Stradling, 29–31.

Page 154: Source of smoke: A detailed study by the Chicago's Association of Commerce in the early 1910s blamed 22 percent of the smoke on locomotives, 29 percent on metallurgical furnaces, 45 percent on high-pressure steam plants, and only 4 percent on domestic fires. Stradling, 78. Eight percent of nation's coal wasted: Stradling, 31.

Pages 154–155: Involvement of engineers: Grinding, in Melosi, 89. Regulators were coal-burning experts, focused on furnace adjustments: Stradling, 105.

Page 155: Oil and gas less than 10 percent: Stradling, 11. Oil running out in fourteen years, oil and gas supplies "ephemeral": Jeffrey, 13–14, 165.

Pages 155–156: Forty-six percent improvement in Pittsburgh's air: Stradling, 76–77.

Page 156, text and footnote: German empire "was built more truly on coal and iron": quoted in Yergin, 544. "The so-called decadence of certain of the European races": Jeffrey, vii. Coal's use to make modern explosives, new medicines, etc.: Jeffrey, 12–13.

Page 157: WWI production spike: Schurr, 74–75. "War meant smoke": Stradling, 138. Severe coal shortage, conservation measures: J. Johnson, 58–77. "Worthy of a Bolshevik Government": J. Johnson, 65–67.

Page 157: Unprecedented labor unrest: Painter, 376–379. Coal battles in the United States more fierce than in Britain: Laslett, 6. Coal bloodier than other U.S. industries: Finley, 121.

Page 157, footnote: Shootings in Pittsburgh, Lattimer, and Colorado (Ludlow): Finley, 120–123; Miller and Sharpless, 230–239; Long, 272–299.

Page 158: Logan and Mingo counties: Finley, 125–129; Coleman, 94–104; Alinsky, 41. Drop in coal use by 1932: J. Johnson, 124–125. Coal as percentage of nation's energy, and drop in total and per capital consumption: Schurr, 36, 64.

Page 158: UMW organizing drive: J. Johnson, 163–166.

Page 159: 1935 break with A.F.of L.: Dubofsky, 160–161.

Page 159: Sudden success of CIO: Dubofsky, 162–204. Lewis's impact on labor movement: Finley, 26–27, 74.

Pages 159–160: Lewis's fame: Dubofsky, 205; Alinsky, 194–195. Lewis hated for wartime strikes, coal wages high in 1949: Finley, 102–106, 116.

Page 160: Statue in Caracas "to honor him as one of the greatest benefactors": Yergin, 543.

Pages 160–161: "couple shovelfuls of coal": Shurick, 36. Coal's shrinking share of energy consumption: Schurr, 36.

Chapter 7: Invisible Power

Page 165: Sherco consumes three-quarters of net consumption in the United States in 1850 (8.5 million tons): Schurr, 70.

Page 166: Nine out of ten tons of American coal used to generate electricity: Energy Information Administration (EIA), *Annual Energy Review 2000,* table 7.3.

Page 167: Brief history of 1970 Clean Air Act: Bryner, 98–101.

Pages 167–168: London fog of 1952, coal banned: Wilson and Spengler, 4–6. Four thousand deaths: Brimblecombe, 112. Traffic slowed, opera cancelled, visibility eleven inches, fifty bodies in one city park: Carr, 46–48.

Page 168: SO_2 emissions reaching an all-time high, doubling every decade: Environmental Protection Agency (EPA), *National Air Pollutant Emission Trends, 1900–1995*, October 1996, p. 3, Figure 1, and p. 8. Greatest sources of SO_2 were coal plants: EPA, *National Air Pollutant Emission Trends, 1900–1998*, March 2000, table 3.7.

Page 169: Scandinavia, Adirondacks, green hair: Anne LaBastille, "Acid Rain: How Great a Menace?" *National Geographic* 160, no. 5 (November 1981): 652–680.

Page 169: Science of acid rain: Park, chapters 3–6. "Green as a birch in spring": LaBastille, *supra,* 675.

Page 170: Industrial response to acid rain science: "Conoco [owner of Consolidation Coal] Chairman Says More Acid Rain Research Needed Before Any Regulatory Action Taken," in *Air/Water Pollution Report,* 5 May 1980, 176; "Conflicting Testimony Presented to Senate Committee in Further Hearings on Acid Rain," in *Air/Water Pollution Report,* 2 June 1980, 214. "A campaign of misleading publicity": "National Coal Association Begins to Stoke Up Debate on Next Year's Clean Air Act Reauthorization," in *Air/Water Pollution Report,* 3 November 1980, 433. Acid Rain Program requirements: EPA, *Progress Report on the EPA Acid Rain Program,* November 1999, 6; EPA, *National Air Pollutant Emissions Trends, 1900–1998,* March 2000.

Page 170: SO_2 costs so much less than predicted: The Electric Power Research Institute predicted costs of the Acid Rain Program would be $4–23 billion/year and the National Wildlife Federation predicted costs of $2.4 billion/year. The U.S. Energy Information Administration found in 1995 that actual costs were about $836 million/year, and although they will rise in years ahead, they will remain well below early predictions: Don Munton, "Dispelling the Myths of the Acid Rain Story," *Environment* 40, no. 6 (July/August, 1998): 4–34; Richard Kerr, "Acid Rain Control: Success on the Cheap," *Science* 282 (November 6, 1998): 1024.

Pages 170–171: SO_2 cuts of 1990 may not be enough: C. T. Driscoll et al., *Acid Rain Revisited: Advances in Scientific Understanding Since the Passage of the Clean Air Act Amendments,* Hubbard Brook Research Foundation, Science Links Publication, report 1, no. 1, 2001. Rain still occasionally ten times more acidic than normal: Clean Air Task Force, *Unfinished Business: Why the Acid Rain Problem Is Not Solved,* a Clear the Air report, October 2001.

Page 171: Canadian lakes continue to acidify: *Environment Canada,* 1996, annual report of the Federal-Provincial Agreements for Eastern Canada Acid Rain Program, July 1997.

Page 171: Environmentalists point to studies: Clean Air Task Force, *Unfinished Business, supra.* Will take up to a quarter century for some ecosystems to recover: Driscoll et al., *supra.*

Pages 171–172: Miles of visibility lost, main cause is sulfates: EPA, *Latest Findings on National Air Quality: 2000 Status and Trends,* September 2001, 18–19. Haze problem at Grand Canyon: EPA, *Project MOHAVE Final Report,* 19 March 1999. Visibility benefits alone justify pollution controls: Dallas Burtraw, *Cost Savings, Market Performance, and Economic Benefits of the U.S. Acid Rain Program,* Resources for the Future, September 1998.

Page 172: Eighty-one million live in ozone nonattainment, health effects, slow improvement: EPA, *Latest Findings on National Air Quality, 2000 Status and Trends,* September 2001, 5, 7–9. Ozone associated with 10–20 percent of summertime hospital admissions, and children most at risk: EPA, *Fact Sheet, EPA's Revised Ozone Standard,* 17 July 1997, 3–4.

Pages 172–173: Coal contributed nearly a quarter of the nation's NO_x emissions in 1998, more than cars and light-duty trucks (pickups, vans, and SUVs) combined, based on 1998 data: EPA, *National Air Pollutant Emission Trends, 1900–1998,* March 2000, table 3-2. NO_x emissions contribute to eutrophication of waters such as Chesapeake: EPA, *Latest Findings on National Air Quality, 2000 Status and Trends,* September 2001, 6.

Page 173: Mercury emissions from coal plants estimated to be 32.6 percent of total: EPA, *Mercury Study Report to Congress,* vol. 1, December 1997, Executive Summary, table 3-1. Sixty thousand babies born at risk per year: National Research Council, *Toxicological Effects of Methylmercury,* National Academy Press, 2000, 325.

Page 173, footnote: Other toxics, concern over dioxins and arsenic: EPA, *Study of Hazardous Air Pollutant Emissions from Electric Utility Steam Generating Units: Final Report to Congress,* vol. 1, February 1998, Executive Summary, 27; and EPA, *Fact Sheet: EPA to Regulate Mercury and Other Air Toxics Emissions from Coal- and Oil-Fired Power Plants,* 14 December 2000.

Page 175: Deaths of 60,000 to 70,000: Wilson and Spengler, 210; J. Schwartz, "Harvesting and Long Term Exposure Effects in the Relation Between Air Pollution and Mortality," *American Journal of Epidemiology* 151, no. 5 (2000): 440–448. EPA found that reducing particulates to its new standard would save 15,000 lives/year: Carol Browner, EPA Administrator, oral testimony before the U.S. Senate Committee on Agriculture, Nutrition and Forestry, 22 July 1997.

Page 175: Particulates from power plants killing over 30,000 per year: Clean Air Task Force, *Death, Disease and Dirty Power: Mortality and Health Damage Due to Air Pollution from Power Plants,* October 2000, 3, 5.

Page 175: Comparative traffic, homicide, and HIV deaths: Centers for Disease Control, "Deaths: Preliminary Data for 2000," *National Vital Statistics Reports* 19, no. 12 (9 October 2001): table 2. Deaths from black lung: Centers for Disease Control, *Work Related Lung Disease (WoRLD) Surveillance Report 1999,* table 2.1. Coal mining injuries: Bureau of Labor Statistics, *2000 Census of Fatal Occupational Injuries Data,* table A–1.

Page 176: Coal industry views reports blaming coal for health problems with sense of injustice, burning three times more coal than in 1970 while SO_2 and NO_x levels decline: See remarks of Jack. N. Gerard, CEO National Mining Association, "If It Moves in a Circle, It's Probably Spin: Persuasion and Power, Coal and Consent," to the Clean Coal and Power Conference, 19 November 2001.

Pages 177–178: Coal provides just over half of U.S. electricity, with huge regional variations: Energy Information Administration (EIA), *Electric Power Annual 2000,* vol. 1, p. 7 (52 percent), and table 7 (percent of electricity generated by coal by state).

Page 178: Scrubber technology: Noyes, 53–55.

Page 179: Shift of production from east to west: EIA, *Annual Energy Review 2000*, table 7.2.

Page 179: Nearly two-thirds of American coal from surface mines: EIA, *Annual Energy Review 2000*, table 7.2. Coal really cheap: EIA, *The U.S. Coal Industry in the 1990s: Low Prices and Record Production*, September 1999.

Pages 179–180: Mines open in 1976: EIA, *Changing Structure of the U.S. Coal Industry: An Update*, July 1993, table 1. Job losses in twentieth century: EIA, *Coal Data: A Reference*, February 1995, table 18. Mines and employment in 2000: EIA, *Coal Industry Annual 2000*, Preliminary Tables, table 2 and table 40. UMW membership and production: *Coal Industry Annual 2000*, Preliminary Tables, table 11 and table 46.

Page 180: Mountaintop removal and controversy: "Shear Madness," *U.S. News and World Report* (11 August 1997); "In West Virginia: A Coal-Digging Rift," *Boston Sunday Globe*, 4 March 2001; "King Coal: A West Virginia Case Study in Arrogance, an Activists Account," *TomPaine.com*, 22 August 2000. Golf course at reclaimed site: "Arch Coal Honored with West Virginia's Top Mine Reclamation Award for Championship Golf Course Project," PR Newswire-FirstCall, 16 January 2002.

Pages 180–181: EPA enforcement action: EPA, remarks of Carol Browner at Clean Air Enforcement Press Conference, 3 November 1999. Bush has raised concerns: "EPA Seeks to Narrow Pollution Initiative: Utilities Fight Clinton Rules on Coal-Fired Power Plants," *Washington Post*, 8 August 2001; "Justice Department Denies Slowing Actions Against Polluters," *Washington Post*, 10 August 2001.

Page 181: New York and New England complaints of upwind pollution: see, for example, "Clean Air Counterattack," editorial, *New York Times*, 8 November 1999.

Page 181: Health risk thought greatest twenty miles downwind: Jonathan Levy et al., *Estimated Public Health Impacts of Criteria Pollutant Air Emissions from the Salem Harbor and Brayton Point Power Plants*, May 2000, available at Harvard School of Public Health Web site, http://www. hsph.harvard.edu/papers/plant/plant.pdf. Mortality rates by state: Clean Air Task Force, *Death, Disease, and Dirty Power*, *supra*, 22.

Page 182: Studies suggest that even with controls for three pollutants coal would still be economically viable, but carbon dioxide controls would promote dramatic shift away from coal: EIA, *Analysis of Strategies for Reducing Multiple Emissions from Power Plants: Sulfur Dioxide, Nitrogen Oxides, and Carbon Dioxide*, December 2000, xii; EIA, *Reducing Emissions of Sulfur Dioxide, Nitrogen Oxides, and Mercury from Electric Power Plants*, September 2001.

Page 184: CO_2 levels up about a third, highest in millions of years, could go up to two to three times preindustrial level in next century, takes centuries to go back down: Intergovernmental Panel on Climate Change (IPCC), *Climate Change 2001: The Scientific Basis,* Summary for Policymakers, 7, 12, 17.

Pages 184–185, text and footnotes: Contributions of coal, oil, and natural gas to CO_2 emissions: EIA, *Annual Energy Review 2000,* table 12.3. CO_2 emissions of coal, oil, and gas by energy output: EIA, *Greenhouse Gases, Global Climate Change, and Energy.* CO_2 emissions from new gas, old coal, and new coal plants: Linda S. Taylor, "Environmental Issues for Electrical Generation," Minnesota Department of Commerce (paper presented to the Minnesota Environmental Law Institute, Bloomington, Minn., 8 May 2002).

Page 185: 1990s the warmest decade in record, likely warmest in thousand years, retreating glaciers, decline of North Polar ice cap since 1950s: IPCC, *Climate Change 2001: The Scientific Basis,* Summary for Policymakers, 2–4. Plants and animals already shifting: IPCC, *Climate Change 2001: Impacts, Adaptation, and Vulnerability,* Summary for Policymakers, section 2.1. Antarctic ice shelf collapse: Andrew C. Revkin, "Large Ice Shelf in Antarctica Disintegrates at Great Speed," *New York Times,* 20 March 2002. Submarine data shows North Polar ice cap was on average ten feet thick between 1958 to 1976, but on average only six feet thick between 1993 and 1997: William K. Stevens, "Thinning Sea Ice Stokes Debate on Climate Debate," *New York Times,* 17 November 1999.

Pages 185–186: Predicted warming of 1.4 to 5.8 degrees Celsius by 2100: IPCC, *Climate Change 2001: The Scientific Basis,* Summary for Policymakers, 13.

Page 186: Average global temperature during last ice age only 5–6 degrees Celsius colder than today: Houghton, 80.

Page 186: More frequent heat waves and droughts, more flooding, rising sea levels: IPCC, *Climate Change 2001: The Scientific Basis,* Summary for Policymakers, 15. Threat to coastal areas, floods and landslides: IPCC, *Climate Change 2001: Impacts, Adaptation, and Vulnerability,* Summary for Policymakers, sections 3.4–3.6, and chapter 6.4. Smog and water pollution problems, increased risk of forest fires: National Assessment Synthesis Team, *Foundation,* 446–450, 500.

Pages 186–187: Decline of boreal forests and loss of sugar maple, spread of invasive species: National Assessment Synthesis Team, *Foundation,* 502, 507. Significant disruption of ecosystems, problems of fragmented habitat, invasive species, extinction rates increase: IPCC, *Climate Change 2001: Impacts, Adaptation, and Vulnerability,* Technical Summary, section 4.3; Gian-Reto Walther et al., "Ecological Responses to Recent Climate Change," *Nature* 416 (28 March 2002): 389–395.

Page 187: Potential increase in crop and timber yields, lower energy costs, and CO_2 fertilizer effect: IPCC, *Climate Change 2001: Impacts, Adaptation, and Vulnerability,* Summary for Policymakers, sections 2.4 and 2.6., and Technical Summary, section 4.3.

Page 188: Loss of predictability: IPCC, *Climate Change 2001: Impacts, Adaptation, and Vulnerability, 2001,* Technical Summary, section 4.1. Increase in year-to-year variability: IPCC, *Climate Change 2001: The Scientific Basis,* Summary for Policymakers, 13, 16.

Page 188: Greater threats to poorer regions: IPCC, *Climate Change 2001: Impacts, Adaptation, and Vulnerability,* Summary for Policymakers, section 4, and chapters 10 and 11. Climate disruptions as challenge to global security: National Assessment Synthesis Team, *Overview,* 9.

Pages 188–189: National Academy of Sciences report, "recent scientific evidence shows," new thinking is "little known," and greatest risk when climate being forced to change: National Research Council, Committee on Abrupt Climate Change, *Abrupt Climate Change: Inevitable Surprises,* National Academy Press, 2002, 1, 83.

Page 189, footnote: The shut-down of ocean currents would cause "massive changes" but not a new glacial period: National Research Council, *supra,* 109. Shut-down of ocean currents could cause new ice age: William H. Calvin, "The Great Climate Flip-Flop," *Atlantic Monthly* (January 1993).

Page 190: Kyoto Protocol: text available on the official UN Web site at http://unfccc.int/. Stopping the build-up would have meant cutting CO_2 emissions by half with more cuts to come: IPCC, *Second Assessment Synthesis of Scientific-Technical Information Relevant to Interpreting Article 2 of the UN Framework Convention on Climate Change,* 1995, section 4.6. Measures far more ambitious than Kyoto (and global in scope) will be needed to limit build-up to doubling of natural level: IPCC, *Climate Change 2001: Mitigation,* Summary for Policymakers, fig. SPM1.

Pages 190–191: For a detailed account of the sustained and often misleading campaign by coal and other energy companies to minimize climate change fears and influence public policy, see Gelbspan. Sued environmental groups for putting coal "in an unwholesome and unfavorable light": *Western Fuels v. Turning Point Project,* U.S. District Court, District of Wyoming, case no. 00-CV-074-D. The case was thrown out of court on jurisdictional grounds, and Western Fuels chose not to bring it in the proper jurisdiction.

Page 191: Coal and oil among God's greatest gifts to carry out Genesis command, "the arrogance to attempt to intervene": Fred Palmer, speech before CoalTRANS 96, Madrid, Spain, 21 October 1996.

Pages 191–192: "Our world is deficient": from Western Fuels' video, "The Greening of Planet Earth." The recent head of Western Fuels publicly stated: "The greenhouse growers that have commercial greenhouses run their ambient air at 1,000 parts per million CO_2 in the air. I say bring it on. I asked our scientists, I said, give me a study, 1,000 parts per million, what's it look like? Pretty big tomatoes." This quote, and "warm is good, cold is bad":

Fred Palmer, Global Warming Debate, before the Annual Meeting of the Minnesota Rural Electric Association, 23 February 2000, available online at http://www.me3.org.

Page 192, first footnote: CO_2 stimulus limited by other nutrients, protein levels reduced: IPCC, *Climate Change 2001: Impacts, Adaptation, and Vulnerability*, chapter 5.3, boxes 5-3 and 5-4. Invasive species, threat to biodiversity: Stanley D. Smith et al., "Elevated CO_2 Increases Productivity and Invasive Species Success in an Arid Ecosystem," *Nature*, vol. 408 (2 November 2000): 79; "High CO_2 Levels May Give Fast-Growing Trees an Edge," *Science*, vol. 292 (6 April 2001): 36.

Page 192, second footnote: "they don't want to be associated with us": Ned Leonard, Greening Earth Society, discussion with author, 8 February 2002. Palmer continues to hold the same views, but Peabody doesn't necessarily endorse them: Fred Palmer, discussion with author, 28 February 2002.

Page 193: "Albert Gore was trying": quote from Fred Palmer of Peabody Energy, discussion with author, 28 February 2002. Coal industry tripled its campaign contributions: data from the Center for Responsive Politics, http://www.opensecrets.org.

Pages 193–194: Report on West Virginia election, "it was basically a coal-fired victory," and quotes from head of West Virginia Coal Association, William Raney, that industry was receiving "payback" from President Bush, and that "he is appreciative": Tom Hamburger, "A Coal-Fired Crusade Helped Bring Bush A Crucial Victory," *Wall Street Journal*, 13 June 2001.

Pages 194–195: Coal use drops abroad: EIA, *International Energy Outlook 2001*, 67, 72. "King Coal is Back!" and "Another Record Year for Coal," *Coal Leader*, April 2001 and July/August/September 1999. Some ninety new coal plants: National Energy Technology Laboratory, U.S. Department of Energy, *Tracking New Coal-Fired Power Plants; Coal's Resurgence in Electric Power Generation*, 9 January 2002.

Page 195: Plants required to convert by law: EIA, *Coal Data: A Reference*, February 1995, 31. The Clean Air Act was amended to make it easier for plants to convert to coal; see, for example, 42 U.S.C. section 7411(a)(8).

Pages 195–196: Quotes by the former head of the National Mining Association, Richard L. Lawson, "unilateral economic disarmament" and "for the first time in history": *Coal Leader*, January/February 1998, 2.

Page 196: US represents 24 percent of global CO_2 emissions from fossil fuels, and developed nations represent about two-thirds: EIA, *International Energy Annual 2000*, table H1.

Page 197: China leads the list: Condoleezza Rice, Bush's National Security Advisor, recently stated that "a protocol that excepts China and India . . . won't be ratifiable": Jeffrey Kluger, "A Climate of Despair," *Time* 157, no. 14 (9 April 2001): 30. China burns more coal: EIA, *China Country Analysis Brief,* April 2001. China gets about two-thirds of its energy from coal, though exact estimates vary: EIA, *China Country Analysis Brief,* April 2001 (62 percent of primary energy from coal); Zhou Dadi et al., *Developing Countries and Global Climate Change: Electric Power Options in China,* report prepared for the Pew Center on Global Climate Change, May 2000, 1 (70 percent of primary energy). The United States got two-thirds of its energy from coal in 1925: Schurr, 36, table 2.

Chapter 8: A Sort of Black Stone

Page 199: China's coal deposits second to those of the United States: Tregear, 25.

Page 201, footnote: Ten percent of energy investment is foreign: Zhou Dadi et al., *Developing Countries and Global Climate Change: Electric Power Options in China,* report prepared for the Pew Center on Global Climate Change, May 2000, 13 (available at http://www.pewclimate.org).

Page 202: Rise of agriculture, Great Wall at the boundary: Tregear, 2–3.

Page 202: Rainfall variability, millions have died of famine: Tregear, 61, 108. History of flooding: Chen, 149; Cressey, 248. Yu the Great, first dynasty around 2200 B.C., built ditches and canals, danced to reduce the flood waters: Granet, 16, 63, 189. Built dams: Bodde, 37.

Pages 202–203: Water projects major responsibility of state, mandate of heaven depended on keeping water system maintained: Merson, 19, 37.

Pages 203–204: "Throughout this province": Polo, 170–171.

Page 204: Jet-carving history, "ear-piercing ornaments": Needham and Golas, 190–191.

Page 204: Coal burned around third century B.C.: Li and Lu, 88.

Page 205: China's lead in iron production: Robert Hartwell, "Markets, Technology, and the Structure of Enterprise in the Development of the Eleventh-Century Chinese Iron and Steel Industry," *Journal of Economic History* 26 (1966): 29–58. China's early iron industry, and complaints from 120 B.C.: Needham, *Development of Iron and Steel,* 18. Fuel crisis, switch to coke, expansion of industry: Robert Hartwell, "A Cycle of Economic Change in Imperial China: Coal and Iron in Northeast China, 750–1350," *Journal of the Economic an Social History of the Orient* 10, no. 119 (1967): 115–119.

Pages 205–206: Large, coke-fueled enterprises, production and employment figures: Hartwell, "A Cycle of Economic Change," *supra,* 115–123.

Page 206: Kaifeng described, "a multifunctional urban center," population estimates: Hartwell "A Cycle of Economic Change," *supra,* 125–128.

Page 206: Weaponry statistics: Robert Hartwell: "A Revolution in the Chinese Iron and Coal Industries During the Northern Sung, 960–1126 A.D.," in *Journal of Asian Studies* 21, no. 1 (February 1962): 157–158; Hartwell, "Markets, Technology and Structure," *supra*, 38; Merson, 22.

Pages 206–207: Charcoal stampede, and coal use becomes prevalent by early 1100s: Hartwell, "A Cycle of Economic Change," *supra*, 133, 135–141.

Page 207: Threat from north, decline of navy, ban on trade and language: Merson, 75–77. China's turn inward: Fairbank and Goldman, 137–140.

Pages 207–208, text and footnote: Five million Chinese miners in 1996, numbers of mines and miners, small Chinese mines vastly more dangerous: Coal Industry Advisory Board, 30. About 90,000 U.S. coal jobs in 1995: National Mining Association, *Coal Mining and the American Economy*, July 1997. Ten thousand deaths in Chinese mines in 1991: Dorian, 247. Fifty-one deaths in U.S. mines in 1992: *Bureau of Labor Statistics*, table A–3, "Fatal Occupational Injuries by Industry, 1992 and 1993."

Page 210: "Some galleries are twice the height," quote from Kangxi period, 1662–1722: Xu and Wu, 289. Miners lowered on a rope: William H. Shockley, "Notes on the Coal- and Iron-Fields of Southeastern Shansi, China," *Transactions of the American Institute of Mining Engineers* 34 (1904): 841–871, 860. Lantern in pigtail: Alexander Reid, "Chinese Mines and Miners," *Transactions of the Institution of Mining Engineers* 23 (1901–1902): 26–37, 33.

Pages 210–211: Hours of work: Wright, 169–171; Reid, *supra*, 31. "A particularly weird site": Shockley, *supra*, 863.

Page 211: Slave labor in the mines, even into 1900s: Wright, 165. Slave labor, and official kept underground who "bit his finger": Xu and Wu, 303–304. Child and women workers: Wright, 161–163.

Page 211: Lack of engines, use of buckets, coastal cities imported coal from abroad: Wright, 5, 36, 52.

Page 212: Opium war and treaty ports: Fairbank and Goldman, 198–205; Merson, 172. Coal demand in treaty ports, growing interest in China's coal: Wright, 51; Carlson, 14–15.

Pages 212–213: Self-strengthening movement: Merson, 174, Fairbank and Goldman, 217–221. Li Hongzhang points to Britain's coal power, founds Kaiping mines over *feng-shui* and earth dragon objections, "the first successful, large-scale effort": Carlson, 1–16.

Page 213: Beijing kept warm by "sooty and weary camels": Andersson, 23. Rocket of China story, "nothing more was said about mules": Carlson, 18–22.

Pages 213–214: Moreing provides capital, brings over Hoover: Carlson, 54–57, 60; Nash, 97–100.

Page 214: Boxer rebellion, missionaries accused of eating babies, controlling weather, missionaries murdered: Preston, 28–30, 279. "Righteous Harmonious Fists": Burner, 36.

Page 214: Kaiping mines put under British flag, greatest industrial enterprise in China conveyed to Hoover: Nash, 126–159; Carlson, 57–83.

Page 215: Hoover manages the mine: Nash, 160–170; Burner, 39. Kaiping becomes cause célèbre, Chinese "fleeced" and "got themselves fairly had," and continued British control: Carlson, 2, 80–94, 138. Coal-poor Japan needed Manchurian coal: Wang, 27.

Page 216: One of world's richest deposits of coal: Needham and Golas, 186.

Page 216: Loess soils up to 250 feet deep, erosion patterns: Tregear, 26–31; Andersson, 128. No carts in parts of the loess, just pack animals: Gillin, 91. Loess makes transportation hard: Chen, 4; Cressey, 269. Shanxi coal largely unexploited till modern times: Tregear, 190.

Page 217: "The biggest and most ambitious experiment": Karnow, 92. Krushchev's goal to surpass the United States, Mao's goal to surpass Great Britain: MacFarquhar, 2:16–18; Becker, 56. Tapping vast rural population: Karnow, 92; Fairbank and Goldman, 370.

Pages 217–218: Establishment of communes: Fairbank and Goldman, 370; Karnow, 92. Peasants building dams: Karnow, 94, 97; Becker, 77–79; Fairbank and Goldman, 370–371. Small industrial enterprises set up by villages: Gao, 131.

Page 218, text and footnote: "The whole people making steel": Tiewes and Sun, 110. Million furnaces making steel, a hundred million people: Fairbank and Goldman, 371. Two million furnaces making steel: Dorian, 36. Steel targets: MacFarquhar, 2:88–90. Actually making iron, furnaces built in hours: MacFarquhar, 2:114, 116.

Page 218: One hundred thousand coal pits, 20 million miners: Merson, 230. One hundred thousand prisoners at one coal mine: Becker, 187.

Page 219: Six hundred thousand to clean coal, transportation crisis caused by coal shipping, and Mao admits he failed to predict transportation overload: Macfarquhar, 2:168, 129, 89. Hand-pulled carts: Karnow, 101. Ni Yuxian as schoolboy hauling coke: Thurston, 56–57.

Page 219: Countryside stripped bare for fuel: Chang, 222; Gao, 131. Doors removed: Becker, 2.

Pages 219–220: "Cattle droppings": Chang, 222. Steel was largely unusable: Fairbank and Goldman, 371. Later generations would inherit congealed heaps: Karnow, 102; Thurston, 58.

Page 220: Famine killed tens of millions: Becker, 267–274; Fairbank and Goldman, 368; MacFarquhar, 3:4; Gao, 126. Eating leaves, etc.: Becker, 2–4. Economy in shambles, years to recover: Fairbank and Goldman, 372–373. Retreat from Great Leap Forward: MacFarqhuar, 3:13–18.

Pages 220–221: Shift from north to south and back, and coal production, 1949–1989: Lu, 5–6, 14. Cultural Revolution: Fairbank and Goldman, 383–405.

Pages 221–222, text and footnote: Thirty thousand small mines have been closed since 1998: Energy Information Administration (EIA), *China Country Analysis Brief,* April 2001. Fatalities at closed mines: Jonathan E. Sinton and David G. Fridley, "What Goes Up: Recent Trends in China's Energy Consumption," Lawrence Berkeley National Laboratory, 25 February 2000. Eight hundred and eighty-three thousand coal miners laid off, nearly 800,000 more planned: Coal Industry Advisory Board, 47. Communist government urges cartels: "Government plans to set up cartels," *China Daily,* 5 July 2001.

Page 222: Economic growth almost fastest in history, quadrupled per capita income: Fairbank and Goldman, 406.

Page 222: Caves in loess cliffs since Neolithic times: Fairbank and Goldman, 14.

Pages 222–223: Description of *kang,* "sometimes it happens": Needham, *Science and Civilisation in China,* 4:135.

Page 223: Chinese have been cooking with coal since seventh century: Needham and Golas, 194. Coal good for cooking food, except bean curds: Sung, 206.

Pages 223–224: Dissidents in 1979 finally able to point out energy crisis, some unable to cook one meal per day: Lu, 3–4, 8–9.

Page 224: Capacity rise between 1988 and 1999: Zhou et al., *supra, 3.* China cautious so as not to repeat California's mistakes: "Power Reforms Set to Slow," *China Daily,* 24 April 2001.

Pages 224–225: Moratorium on coal plants, focus on distribution: Zhou et al., *supra,* 2. Television and refrigerator ownership in rural China: *The Controversy Over China's Reported Falling Energy Use,* report from the U.S. Embassy, Beijing, August 2001. Chinese still use thirteen times less electricity than the United States, capacity to double by 2015, and coal on track to provide 85 percent of electricity: Zhou et al., *supra,* 1, 22–23.

Page 225, text and footnote: 1998 flooding along Yangzte: "Environmental Neglect Faulted in China's Floods," Associated Press report, *Star Tribune* (Minneapolis), 27 August 1998. Dust storms in Beijing, now more frequent: *Grapes of Wrath in Inner Mongolia,* report from U.S. Embassy, Beijing, May 2001.

Pages 225–226: China dust storms travel to North America: "China Dust Storm Strikes USA," statement by the National Oceanic and Atmospheric Administration, 18 April 2001; "Second Asian Dust Storm Over U.S.," statement by the National Oceanic and Atmospheric Administration, 24 April 2001. Aspen pollution levels: *Grapes of Wrath in Inner Mongolia,* report from U.S. Embassy, Beijing, May, 2001.

Page 226: Five of ten most polluted cities: Coal Industry Advisory Board, 53. China now top in steel production: EIA, *International Energy Outlook, 2001,* 70.

Page 227, text and footnote: Million premature deaths from air pollution: *PRC Air Pollution: How Bad Is It?* report from U.S. Embassy, Beijing, June 1998. One in eight deaths: *The Cost of Environmental Degradation in China,* report by the U.S. Embassy Beijing, December 2000. Ten million suffering from fluorine and arsenic poisoning: Robert B. Finkelmen et al., "Health Impacts of Domestic Coal Use in China," *Proceedings of the National Academy of Sciences* 96, no. 7 (30 March 1999): 3427–3431; Sarah Simpson, "Coal Control: Tackling the Health Dangers of China's Dirty Coal," *Scientific American* (February 2002): 20.

Page 227: China's SO_2 emissions highest in world, acid rain affects 40 percent of land: Zhou et al., "Developing Countries," *supra,* 14. Acid rain more a problem in the south, neutralized in the north: Smil, *China's Environmental Crisis,* 118. Japan suffers acid rain from China's emissions: *If Shanxi Can Do It (Clean Up), Anybody Can,* report from U.S. Embassy Beijing, June 2001.

Pages 227–228: China's admission of problem: *Severe Beijing Air Pollution Information Emerges,* report from U.S. Embassy Beijing, February 1998. Chinese-style campaigns: In 1998, I interviewed an official with China's State Environmental Protection Agency, who described Beijing's closure of tens of thousands of polluting industrial facilities. See also, *If Shanxi Can Do It (Clean Up), Anybody Can,* report from U.S. Embassy Beijing, June 2001.

Page 228: Improvements in Beijing SO_2 levels, and comparison to World Health Organization guidelines: *Ninth Five-Year Plan Environmental Report Card,* report from the U.S. Embassy, March 2001. SO_2 emissions continuing to rise: Zhou et al., *supra,* 15.

Page 229: The Beijing Energy Efficiency Center's Web site is at http://www.beconchina.org. Zhou's fate during Cultural Revolution: Hertsgaard, 188.

Pages 229–230: China using less energy than similar nations, CO_2 emissions half what they would otherwise be: Jeffrey Logan, Aaron Frank, Jianwu Feng, and Indu John, "Climate Action in the United States and China," Pacific Northwest National Laboratory, May 1999, available at http://www.pnl.gov.

Page 230: Reports of 17 percent drop in energy output, and 30 percent drop in coal use: *The Controversy Over China's Reported Falling Energy Use,* report from the U.S. Embassy Beijing, August 2001. A recent analysis suggests that half of a reported 20 percent drop in China's coal use over five years may be due to unreported coal consumption, but that China is still emitting far less CO_2 than earlier forecasts had predicted. "An Update on Recent Energy and Carbon Dioxide Trends in China," Jeffrey Logan, Pacific Northwest National Laboratory, June 2001 (available at http://www.pnl.gov). China's CO_2 emissions may have fallen 6 percent between 1997 and 2000: EIA, *International Energy Annual, 2000,* table H1.

Pages 230–231: Coal use appears to have picked up again: Logan, "An Update," *supra.* China planning to expand electricity capacity using coal, has centuries-worth of coal reserves:

Zhou et al., *supra*, 5, 22. "When we are richer," and "markets surely won't drive this": Zhou Dadi, interview with author, 19 March 2001.

Chapter 9: A Burning Legacy

Pages 236-237: From pesticides to perfume, laughing gas to TNT: Coal has been used as the basic feedstock of an extensive chemical industry whose products include these items and many others, such as nylon, plastics, saccharin, linoleum, sulfa drugs, aspirin, phonographs, DDT, cyanide, moth balls and paint. "Coal," *World Book Encylopedia,* 1961.

Page 237, text and footnote: Carbon sequestration technology is new to scientific lexicon: National Energy Technology Laboratory, U.S. Department of Energy, *Carbon Sequestration Technology Roadmap: Pathways to Sustainable Use of Fossil Energy,* 7 January 2002, 5. Biological processes might be able to temporarily sequester 10–20 percent of fossil fuel carbon emissions: IPCC, *Climate Change 2001: Mitigation,* Technical Summary, 41–42.

Page 238: Geological formations: National Energy Technology Laboratory, *Carbon Sequestration Technology Roadmap, supra,* 13. Limited availability of geological sites: S. H. Kim and J. A. Edmonds, *Potential for Advanced Carbon Capture and Sequestration Technologies in a Climate Constrained World,* report prepared for the U.S. Department of Energy, February 2000, vii. Considering ocean injection: National Energy Technology Laboratory, *Carbon Sequestration Technology Roadmap, supra,* at 18. Impact on deep sea creatures: Brad A. Seibel and Patrick J. Walsh, "Potential Impacts of CO_2 Injection on Deep-Sea Biota," *Science* 94 (October 12, 2001): 319.

Pages 238–239: Lake Nyos tragedy: Marguerite Holloway, "The Killing Lakes," *Scientific American* 283, no. 1 (July 2000): 92–99; "Removing CO_2 from Lake Nyos in Cameroon," *Science* 292 (April 20, 2001): 292, 438.

Page 239: Part of thoroughly redesigned, futuristic coal plants decades away: Federal Energy Technology Center, Department of Energy, *Vision 21 Program Plan: Clean Energy Plants for the 21st Century.*

Pages 241–244: On the hydrogen economy, dropping prices for wind and solar resources, fuel cells, decentralized generation, etc., see, for example: Seth Dunn, "Hydrogen Futures: Toward a Sustainable Energy System," August 2001, Worldwatch Paper 157, available at http://www.worldwatch.org; Christopher Flavin and Seth Dunn, "Reinventing the Energy System," in L. Brown et al.; David Stipp, "The Coming Hydrogen Economy," *Fortune* (November 12, 2001). Jules Verne calling water/hydrogen the "coal of the future": Dunn, *supra*, 6.

Page 245: Coal-bed methane: U.S. Geological Survey, *Coal-Bed Methane: Potential and Concerns,* October 2000.

Pages 247–248: Some future warming from past emissions: IPCC, *Climate Change 2001: The Scientific Basis,* Summary for Policymakers, 13. Our fate is not sealed: the IPCC has set forth a variety of alternative pathways the world might take over the next few decades, with dramatically different outcomes in terms of CO_2 emissions and resulting warming. IPCC, *Climate Change 2001: Mitigation,* chapter 2.

Chapter 10: Tipping Points

Page 251: At its height in early 2007: National Energy Technology Laboratory, *Tracking New Coal-Fired Power Plants: Coal's Resurgence in Electric Power Generation*, 24 January 2007.

Page 252: About four-fifths of projects were blocked: Of the 100 gigawatts of proposed new coal capacity, only about 20 gigawatts were actually added between 2006 and 2013. Energy Information Administration (EIA), "Projected Electric Capacity Additions Are Below Recent Historic Levels," *Today in Energy*, 11 May 2015.

Page 253: Public concern and media attention rose: Robert J. Brulle et al., "Shifting Public Opinion on Climate Change: An Empirical Assessment of Factors Influencing Concern over Climate Change in the U.S., 2002–2010," *Climatic Change*, vol. 114 (September 2012). Announcement that climate change was "unequivocal": Intergovernmental Panel on Climate Change (IPCC), *Climate Change 2007: The Physical Science Basis*, Summary for Policymakers.

Page 253: States with renewable energy standards: Union of Concerned Scientists (UCS), "Renewable Electricity Standards at Work in the States," fact sheet, February 2009. States with energy efficiency programs: Maggie Eldridge et al., *2008 State Energy Efficiency Scorecard*, American Council for an Energy-Efficient Economy, 1 October 2008. States with GHG reduction targets: Center for Climate and Energy Solutions (C2ES), "Climate Action: Greenhouse Gas Emissions Targets," online data. States with carbon-reduction plans: C2ES, "Climate Action: Climate Action Plans," online data.

Page 254: We agreed on a plan: Minnesota Climate Change Advisory Group, *Final Report: A Report to the Minnesota Legislature*, April 2008.

Page 254: States setting up their own cap-and-trade markets: Barbara Freese, Steve Clemmer, and Alan Nogee, *Coal Power in a Warming World: A Sensible Transition to Cleaner Energy Options*, UCS, October 2008, chapter 9.

Page 254: Waxman-Markey: Formally called the American Clean Energy and Security Act of 2009 (H.R. 2464), it passed by a vote of 219 to 212.

Page 255: "Deceitful, hysterical, out-of-control rampage": Robert E. Murray (CEO Murray Energy), "Toward a Clean Energy Future: Energy Policy and Climate Change on

Public Lands," testimony before Energy and Mineral Resources Subcommittee, Committee on Natural Resources, U.S. House of Representatives, 20 March 2007.

Page 255: Climate denial had taken root among conservatives by 2006: Jocelyn Kiley, "Ideological Divide over Global Warming as Wide as Ever," Pew Research Center, 6 June 2015.

Page 256: Pawlenty, a "stupid" mistake: Alex Seitz-Wald, "Pawlenty: 'Every One of Us' Running for President Has Flip-Flopped on Climate Change," ThinkProgress, 29 March 2011.

Page 256: Blankenship, global warming "pure make-believe," etc.: Speech to Friends of America Labor Day Festival, at mountaintop mine site near Logan, West Virginia, 9 September 2009 (online video posted by Fluxview.com). Alpha's covert funding of climate denial and scientist harassment: Lee Fang, "Giant Coal Company Bankruptcy Reveals Secret Ties to Climate Denial, GOP Dark Money Groups," Intercept, 25 August 2015; Lee Fang, "Attorney Hounding Climate Scientists Is Covertly Funded by Coal Industry," Intercept, 25 August 2015.

Page 256, footnote: Blankenship indictment: Ken Ward, Jr., "Longtime Massey Energy CEO Don Blankenship Indicted," Charleston Gazette-Mail, 13 November 2014.

Page 257: Koch foundations bankroll groups dismissing climate threat: Robert J. Brulle, "Institutionalizing Delay: Foundation Funding and the Creation of U.S. Climate Change Counter-Movement Organizations," Climatic Change, vol. 122 (February 2014); Greenpeace, Koch Industries Secretly Funding the Climate Denial Machine, March 2010; Lee Fang, "From Promoting Acid Rain to Climate Denial: Over 20 Years of David Koch's Polluter Front Groups," ClimateProgress, 1 April 2010. Rallies against cap-and-trade: Schulman, 274; Russell Gold, "Astroturfing the Climate Bill," Wall Street Journal Blog, 17 August 2009. Americans for Prosperity support for Tea Party: Jane Mayer, "Covert Operations: The Billionaire Brothers Who Are Waging a War Against Obama," New Yorker, 30 August 2010; Schulman, 275.

Page 257: Undoing policies put in place years earlier: Steven Mufson and Tom Hamburger, "Ohio Governor Signs Bill Freezing Renewable-Energy Standards," Washington Post, 13 June 2014. Cap-and-trade operating successfully: Peter Shattuck and Jordan Stutt, The Regional Greenhouse Gas Initiative: A Model Program for the Power Sector, Acadia Center, July 2015; Katherine Hsia-Kiung and Erica Morehouse, Carbon Market California: A Comprehensive Analysis of the Golden State's Cap-and-Trade Program, Year Two: 2014, Environmental Defense Fund, 2014.

Page 257: Public opinion change between 2006 and 2009, by party: Pew Research Center, "Fewer Americans See Solid Evidence of Global Warming," 22 October 2009.

Page 258: Statements by major scientific organization, 97 percent of climate scientists, consensus comparable to tobacco: American Association for the Advancement of Science (AAAS), *What We Know: The Reality, Risks, and Response to Climate Change*, 2014. Poll finds one in ten Americans understand consensus: Anthony Leiserowitz, Geoff Feinberg, and Seth Rosenthal, *Climate Change in the American Mind*, Yale Project on Climate Change Communication and the George Mason University Center for Climate Change Communication, March 2015.

Page 258, footnote: "Climategate" scientists exonerated: UCS, "Debunking Misinformation About Stolen Climate Emails in the 'Climategate' Manufactured Controversy," online article, 2010.

Page 259: Emissions returning to worst-case scenario: John Abraham, "'Very Worried' About Escalating Emissions? You Should Be," *Conversation*, 2 June 2011.

Page 259: "We will use it all": Gregory H. Boyce (CEO Peabody Energy), testimony before Select Committee for Energy Independence and Global Warming, U.S. House of Representatives, 14 April 2010.

Page 259: Widespread dust bowl conditions: Joseph Romm, "The Next Dust Bowl," *Science*, vol. 478 (27 October 2011). Mass extinctions: Kolbert; Mark C. Urban, "Accelerating Extinction Risk from Climate Change," *Science*, vol. 348 (1 May 2015). Ocean acidification: The current rate of ocean acidification is likely the fastest in 300 million years. AAAS, *supra*.

Page 260: IPCC predicts up to one meter sea level rise by 2100 for high-emissions scenario: IPCC, *Climate Change 2013: The Physical Science Basis*, Summary for Policymakers, 25. 150 million people live within a meter of high tide: S. Jevreheva, A. Grinsted, and J. C. Moore, "Upper Limit for Sea Level Projections by 2100," *Environmental Research Letters*, vol. 9 (10 October 2014). Recent analysis suggests upper limit of 1.8 meters by 2100: Jevreheva et al.

Page 260: Unstoppable processes driving ice sheet collapse: IPCC, *Climate Change 2013: The Physical Science Basis*, chapter 13; National Science Foundation, "Global Sea Level Likely to Rise as Much as 70 Feet in Future Generations," press release, 19 March 2012. Part of West Antarctic Ice Sheet past "point of inevitable collapse": NASA, "Decline of West Antarctic Glaciers Appears Irreversible," news release, 16 May 2014.

Page 260: Bombshell of a study: James Hansen et al. "Ice Melt, Sea Level Rise and Superstorms: Evidence from Paleoclimate Data, Climate Modeling, and Modern Observations That 2 Degrees C Global Warming Is Highly Dangerous," *Atmospheric Chemistry and Physics* (23 July 2015).

Page 261: "Headed for unknown territory": Royal Society and National Academy of Sciences, *Climate Change Evidence and Causes: An Overview from the Royal Society and the US National Academy of Sciences*, 2014.

Page 261: Goal of limiting warming: The long-term goal of limiting warming to 2 degrees Celsius and the decision to further consider a goal of 1.5 degrees were agreed to in the Cancun Agreements, adopted at the Sixteenth Conference of the Parties to the Framework Convention on Climate Change, in Cancun, Mexico, 2010.

Page 261: Two degrees Celsius highly dangerous: Hansen et al, *supra*. Greater dangers beyond two degrees Celsius: Subsidiary Body for Scientific and Technological Advice, "Report on the Structured Expert Dialogue on the 2013–2015 Review," UN Framework Convention on Climate Change, June 2015; World Bank, *Turn Down the Heat: Why a 4°C Warmer World Must Be Avoided*, November 2012.

Page 262: Exceeding the world's carbon budget by 2040: International Energy Agency (IEA), *Energy and Climate Change: World Energy Outlook Special Report*, 2015, 35. Different assumptions about future emissions, the Earth's level of sensitivity to rising emissions, and other factors yield somewhat different estimates of when we could exceed our carbon budget. For example, some scientists estimate we could exceed our budget by as soon as 2036 (Michael E. Mann, "Earth Will Cross the Climate Danger Threshold by 2036," *Scientific American*, 18 March 2014). Others estimate we would exceed it in 2044 (P. Friedlingstein et al., "Persistent Growth of CO_2 Emissions and Implications for Reaching Climate Targets," *Nature Geoscience*, vol. 7, 2014, 705–715).

Page 262: An influential study: Carbon Tracker Initiative, *Unburnable Carbon—Are the World's Financial Markets Carrying a Carbon Bubble?*, 2011; see also Carbon Tracker Initiative and Grantham Research Institute on Climate Change and the Environment, *Unburnable Carbon 2013: Wasted Capital and Stranded Assets*, December 2013. Later study finds that 92 to 95 percent of U.S. coal needs to stay buried: Christophe McGlade and Paul Ekins, "The Geographical Distribution of Fossil Fuels Unused When Limiting Global Warming to 2 Degrees C," *Nature*, vol. 517 (January 2015).

Page 263: Bill McKibben publicized concept of unburnable carbon: Bill McKibben, "Global Warming's Terrifying New Math: Three Simple Numbers That Add Up to Global Catastrophe—and That Make Clear Who the Real Enemy Is," *Rolling Stone*, 19 July 2012.

Page 263: Power sector emissions dropped 15 percent: EIA, *Monthly Energy Review*, July 2015, table 12.6. Shift from coal to gas and renewables: EIA, *Monthly Energy Review*, July 2015, table 7.2a. Total greenhouse gas emissions down 9 percent: Environmental Protection Agency (EPA), *U.S. Greenhouse Gas Inventory Report: 1990–2013*, 2015. Coal's falling share of U.S. power market: EIA, *Monthly Energy Review*, July 2015, table 7.2a.

Page 264: "Long-term supercycle for coal": Peabody news release, "Peabody Energy Targets Significant Value Creation amid Long-Term Supercycle for Coal," *PRNewswire*, 17 June 2010.

Page 264: "The insane, regal administration of King Obama": Tim Loh, "A Provocateur Sees Profits in Coal's Long, Slow Death," *BloombergBusiness*, 10 December 2014.

Page 265: Polls show widespread support for coal regulations: Pew Research Center, "How Americans View the Top Energy and Environmental Issues," 15 January 2015; League of Conservation Voters, "LCV Releases New Polling on Clean Power Plan," 27 August 2015.

Page 265: New EPA rules on particulates, mercury, and other issues: For a discussion of EPA rules addressing particulate pollution (Cross-State Air Pollution Rule), mercury (Mercury and Air Toxics Standards), coal ash, cooling water, and carbon, see Barbara Freese, Steve Clemmer, Claudio Martinez, and Alan Nogee, *A Risky Proposition: The Financial Hazards of New Investments in Coal Plants,* UCS, March 2011. Five hundred mountaintops in Appalachia: Ross Geredian, *Post-Mountaintop Removal Reclamation of Mountain Summits for Economic Development in Appalachia*, report prepared for the Natural Resources Defense Council, 7 December 2009.

Page 265–266: Supreme Court ruling on CO_2: *Massachusetts v. Environmental Protection Agency*, 549 U.S. 497 (2007). EPA rules on carbon emissions from coal plants: EPA released these rules, informally titled the Carbon Pollution Standards for New, Modified and Reconstructed Power Plants and the Clean Power Plan for Existing Power Plants, in final form on 3 August 2015.

Page 267: McConnell urges governors not to submit plans: Coral Davenport, "McConnell Urges States to Help Thwart Obama's 'War on Coal,'" *New York Times*, 19 March 2015.

Page 267: Coal power capacity to shrink under Clean Power Plan: EIA, *Analysis of the Impacts of the Clean Power Plan*, May 2015. Same model predicted coal expansion: EIA, *Annual Energy Outlook 2006 with Projections to 2030*, February 2006.

Page 268: Shift from coal has generated jobs: Drew Haerer and Lincoln Pratson, "Employment Trends in the U.S. Electricity Sector, 2008–2012," *Energy Policy*, vol. 82 (July 2015). Solar employment higher than coal: Solar Foundation, "Solar Industry Creating Jobs Nearly 20 Times Faster Than Overall U.S. Economy," press release, 15 January 2015. Coal employment falling again: Paul Krugman, "The War on Coal Already Happened," *New York Times*, 7 June 2014. Central Appalachian coal seams being depleted: Rory McIlmoil et al., *The Continuing Decline in Demand for Central Appalachian Coal: Market and Regulatory Influences*, 14 May 2013. Lost over 14,000 jobs: Dave Mistich, "Central Appalachia, Southern West Virginia 'Ground Zero' for Recent Coal Mine Layoffs," West

Virginia Public Broadcasting and SNL Energy, 17 June 2015. Billions for miners and communities: President's Budget, Fiscal Year 2016, "Investing in Coal Communities, Workers and Technology: The Power+ Plan," fact sheet. Local governments urging congressional approval: Bill Estep, "Local Governments in Eastern Ky. Endorse $1 Billion Obama Plan to Reclaim Coal Land, Create Jobs," *Lexington Herald Leader*, 18 August 2015.

Page 268: "Black is the new green": Boyce testimony, *supra.*

Page 269: CCS embraced by IPCC and others: IPCC, *IPCC Special Report on Carbon Dioxide Capture and Storage*, 2005; Massachusetts Institute of Technology, *The Future of Coal: Options for a Carbon-Constrained World*, 2007.

Page 269: I coauthored a report: Barbara Freese, Steve Clemmer, and Alan Nogee, *Coal Power in a Warming World: A Sensible Transition to Cleaner Energy Options*, UCS, 2008.

Page 269: CCS increases cost of energy: Interagency Task Force on Carbon Capture and Storage, *Report of the Interagency Task Force on Carbon Capture and Storage*, 2010.

Page 270: Projects were scrapped: Ken Wells and Benjamin Elgin, "Carbon Capture Hopes Dim as AEP Says It Got Burned at Coal Plant," *BloombergBusiness*, 20 July 2011. FutureGen scrapped: Jim Snyder and Mark Drajem, "FutureGen's Demise Shows Carbon Capture for Coal Faces Woes," *BloombergBusiness*, 5 February 2015.

Page 270: Saskatchewan plant with CCS: Peter Danko, "World's First Full-Scale 'Clean' Coal Plant Opens in Canada," *National Geographic*, 2 October 2014. Mississippi CCS plant delays and cost overruns: Sonal Patel, "Kemper County IGCC Project Costs Soar to $6.1B," *Power Magazine*, 29 October 2014. CCS would only marginally increase the amount of coal we could burn: Carbon Tracker Initiative and Grantham Research Institute on Climate Change and the Environment, *supra.*

Page 270: Peabody seeks additional subsidies for CCS: Fredrick D. Palmer of Peabody Energy chairs the coal policy committee of a federal advisory committee calling for additional subsidies in support of CCS. National Coal Council, *Fossil Forward—Revitalizing CCS: Bringing Scale and Speed to CCS Deployment*, January 2015. "Non-existent harm" and "benign gas": Peabody Energy, "Comments to the Council on Environmental Quality, Re: Revised Draft Guidance for Federal Departments and Agencies Consideration of Greenhouse Gas Emissions and the Effects of Climate Change in NEPA Reviews," 24 March 2015. "By orders of magnitude" and "no limit . . . to these benefits": Roger Bezdek, direct testimony, In the Matter of Further Investigation into Environmental and Socioeconomic Costs, before the MN Public Utilities Commission (Docket. No. E-999-CI-14-643), 1 June 2015.

Page 271: Expectations for growth of wind and solar surpassed: Year 2000 projections from IEA, *World Energy Outlook 2000*, table 3.9. Actual wind and solar installed capacity data for 2010 and 2014 from International Renewable Energy Agency, Renewable Energy Capacity Statistics, 2015.

Page 271–272: From 5 to 30 percent of global power by 2040: Bloomberg New Energy Finance, *New Energy Outlook 2015*. Almost half of new capacity in 2014 was renewable: IEA, *Energy and Climate Change*, 2015.

Page 272: Price of solar panels fell 99 percent: Zachary Shahan, "Solar Power's Massive Price Drop (Graph)," *CleanTechnica*, 24 May 2013. Other costs of solar business keep falling: Galen Barbose and Naïm Darghouth, *Tracking the Sun VIII: The Installed Price of Residential and Non-residential Photovoltaic Systems in the United States*, 2015. Wind costs fell 58 percent between 2009 and 2014, greater grid parity: Lazard, *Lazard's Levelized Cost of Energy Analysis—Version 8.0*, September 2014.

Page 273: Gigafactory to double global lithium battery production: Reuters, "Tesla to Raise $1.6 Billion to Build Battery Factory," *New York Times*, 26 February 2014. "Staggeringly gigantic": Jeff McMahon, "Elon Musk: Tesla Powerpack Doesn't Need Renewables, Battery Market 'Staggeringly Gigantic,'" *Forbes*, 5 August 2015.

Page 274: Utilities are taking notice: Peter Kind, *Disruptive Challenges: Financial Implications and Strategic Responses to a Changing Retail Electric Business*, report prepared for Edison Electric Institute, January 2013. Barclays downgrades bonds: Michael Aneiro, "Barclays Downgrades Electric Utility Bonds, Sees Viable Solar Competition," *Barrons*, 23 May 2014.

Page 274: Nuclear plants provide 19 percent of U.S. power and 11 percent globally: EIA, *Annual Energy Outlook 2015*, April 2015; IEA/Nuclear Energy Agency, *Technology Roadmap: Nuclear Energy*, 2015. Costs of nuclear power in France: Arnulf Grubler, "The Costs of French Nuclear Scale-Up: A Case of Negative Learning by Doing," *Energy Policy*, vol. 28 (September 2010). Nuclear power's share to rise to 17 percent by 2050: IEA/Nuclear Energy Agency, *supra*. Radical nuclear redesigns currently being pursued: Bryan Walsh, "Amid Economic and Safety Concerns, Nuclear Advocates Pin Their Hopes on New Designs," *Ecocentric*, 5 August 2013.

Page 274, footnote: Shortcomings of fuel cells, most hydrogen made from natural gas: Joe Romm, "Tesla Trumps Toyota: Why Hydrogen Cars Can't Compete with Pure Electric Cars," *ClimateProgress*, 5 August 2014.

Page 275: Official estimates of methane leakage way too low: Oliver Schneising et al., "Remote Sensing of Fugitive Methane Emissions from Oil and Gas Production in North American Tight Geologic Formations," *Earth's Future*, vol. 2 (October 2014); Mark Golden, "America's Natural Gas System Is Leaky and in Need of a Fix, New Study Finds," *Stanford Report*, 13 February 2014. EPA has found some water impacts: EPA's draft study of the impact of fracking on drinking water finds that there are no "widespread, systemic impacts on drinking water," but it identifies potential mechanisms through which water could be impacted. EPA found only a relatively few examples of such impacts, but it goes on to say that this finding might be due to data limitations. EPA, *Hydraulic Fracturing Drinking Water Assessment (Draft)*, executive summary, June 2015. Sudden increase in earthquakes in Oklahoma: Richard D. Andrews and Austin Holland, "Summary Statement on Oklahoma Seismicity," Oklahoma Geological Survey, 21 April 2015.

Page 276: Cheap gas delays deployment of renewables: J. Deyette et al., *The Natural Gas Gamble: A Risky Bet on America's Clean Energy Future*, UCS, 2015; Energy Modeling Forum, *Changing the Game? Emissions and Market Implications of New Natural Gas Supplies*, Stanford, September 2013. Regulations starting to emerge: EPA proposed rules to limit methane emissions from new fracked oil and gas wells on August 18, 2015.

Page 276: Energy efficiency cheap way to cut carbon: Lawrence Berkeley National Laboratory, "Why Is Energy Efficiency the Most Abundant, Cheapest Way to Reduce Greenhouse Gas (GHG) Emissions," online article, accessed September 2015.

Page 277: Coal burning roughly tripled after 2000: There was a decline in coal consumption in China in the late 1990s, as I mention in chapter 8, but current data suggests it was only a 5.7 percent drop, not 10 percent, and it quickly reversed. EIA, *International Energy Statistics*, Total Coal Consumption, China (database). Coal is source of more carbon pollution than oil: In 2000, coal was the source of 38 percent of global energy-related CO_2 emissions, and oil was the source of 42 percent. By 2014, coal was the source of 44 percent of these emissions, while oil's share fell to 35 percent. IEA, *Energy and Climate Change*, 2015.

Page 277: One out of every two tons burned in China: IEA, *Coal Information*, 2015. China world's top greenhouse gas emitter by far: IEA, *Energy and Climate Change*, 28. United States still top greenhouse gas emitter counting historic emissions: Mengpin Ge, Johannes Friedrich, and Thomas Damassa, "6 Graphs Explain the World's Top 10 Emitters," World Resources Institute, 25 November 2014. Coal supplies two-thirds of China's energy: EIA, "China," International Energy Data and Analysis, 2015. Coal supplies 18 percent of U.S. energy: EIA, *Annual Energy Outlook 2015*.

Page 277–278: New export terminals blocked: Katherine Bagley, "Losing Streak Continues for U.S. Coal Export Terminals," *InsideClimate News*, 12 January 2015.

Page 278: Chinese officials start talking about limiting coal use: David Winning, "China's Coal Crisis," *Wall Street Journal*, 16 November 2010. Air pollution kills 1.6 million Chinese yearly: Robert A. Rhode and Richard A. Muller, "Air Pollution in China: Mapping of Concentrations and Sources," *PLoS One*, 2015. Chinese war on pollution: Reuters, "China to 'Declare War' on Pollution, Premier Says," 4 March 2014.

Page 278–279: Banning of *Under the Dome*: Daniel K. Gardner, "China's 'Silent Spring' Moment? Why 'Under the Dome' Found a Ready Audience in China," *New York Times*, 18 March 2015.

Page 279: U.S.-China climate deal: White House, "The U.S. and China Just Announced New Actions to Reduce Carbon Pollution," 12 November 2014.

Page 280: Chinese coal use to peak by 2020: Edward Wong, "In Step to Lower Carbon Emissions, China Will Place a Limit on Coal Use in 2020," *New York Times*, 14 November 2014. Coal use flat or down in 2014 and speculation it might have peaked: Edward Wong and Chris Buckley, "Fading Coal Industry in China May Offer Chance to Aid Climate," *New York Times*, 21 September 2015; EIA, "Recent Statistical Revisions Suggest Higher Historical Coal Consumption in China," 16 September 2015. Coal use falling in early 2015: Lauri Myllyvirta, "China Coal Use Falls: CO_2 Reduction This Year Could Equal UK Total Emissions over Same Period," *EnergyDesk*, Greenpeace, 14 May 2015. Officials increase earlier estimates of Chinese coal use: Chris Buckley, "China Burns Much More Coal Than Reported, Complicating Climate Talks," *New York Times*, 3 November 2015.

Page 280: China restructuring its economy: Fergus Green and Nicholas Stern, *China's "New Normal": Structural Change, Better Growth, and Peak Emissions*, Grantham Research Institute on Climate Change and the Environment and Centre for Climate Change and Economics, June 2015.

Page 281: China has most wind power installed: Global Wind Energy Council, *Global Wind Energy Report 2014: Annual Market Update*, March 2015. China is largest market for solar panels: Katie Fehrenbacher, "China Is Utterly and Totally Dominating Solar Panels," *Fortune*, 18 June 2015. China leads world in nuclear reactor construction: IEA, *Energy and Climate Change*, 2015. China increasing fracking: Sara Sjolin, "China's Shale Ambition: 23 Times the Output in 5 Years," *MarketWatch*, 12 February 2015. China adopting national cap-and-trade: Steven Mufson, "China to Adopt Cap and Trade System to Limit Carbon Emissions," *Washington Post*, 24 September 2015. China considering carbon tax: Brian Palmer, "Coal-tural Revolution," *onEarth*, 2014.

Page 282: McConnell promises to "go to war" with Obama: James Carroll, "McConnell Vows War with Obama over Coal," *Courier Journal*, 14 November 2014. "Our international partners should proceed with caution": Office of Mitch McConnell, "McConnell Statement on Obama Administration International Climate Plan," press release, 31 March 2015. McConnell staff visits foreign officials: Jean Chemnick, "Capitol Hill Messaging Foreign Embassies Ahead of Paris Talks," *E&E News*, 25 August 2015.

Page 283: Russian INDC ambiguously worded: Sophie Yeo and Simon Evans, "Ambiguous Russian Climate Pledge Mystifies Many," Carbonbrief.org, 1 April 2015; IEA, *Energy and Climate Change*, 2015.

Page 283: EU pledge to cut emissions: Arthur Neslen, "EU Leaders Agree to Cut Greenhouse Gas Emissions by 40% by 2030," *Guardian*, 23 October 2014.

Page 283: Analysts find low global price tags given climate benefits: Joe Romm, "It's Not Too Late to Stop Climate Change, and It'll be Super-Cheap," *ClimateProgress*, 29 January 2015; IPCC, *Climate Change 2014: Mitigation of Climate Change*, Summary for Policymakers, table SPM.2; IEA, *World Energy Investment Outlook*, 2014. Citigroup study and investments: Citigroup, *Energy Darwinism II: Why a Low Carbon Future Doesn't Have to Cost the Earth*, August 2015.

Page 284: Based on IPCC scenarios that depend on negative emissions: IPCC, *Climate Change 2014*, Summary for Policymakers, 12; Sabine Fuss et al., "Betting on Negative Emissions," *Nature Climate Change*, 21 September 2014; Oliver Geden, "Climate Advisers Must Maintain Integrity," *Nature*, vol. 521 (6 May 2015); David Roberts, "The Awful Truth About Climate Change No One Wants to Admit," *Vox*, 15 May 2015.

Page 285: More access to mobile phones than toilets: United Nations, "Deputy UN Chief Calls for Urgent Action to Tackle Global Sanitation Crisis," 21 March 2013.

Page 286: Institutions divesting: Emma Howard, "The Rise and Rise of the Fossil Fuel Divestment Movement," *Guardian*, 19 May 2015; Melanie Mason, "Bill to Divest State's Public Pensions from Coal Heads to Gov. Jerry Brown," *LA Times*, 2 September 2015. Bill McKibben quote: David Gelles, "Fossil Fuel Divestment Movement Harnesses the Power of Shame," *New York Times*, 13 June 2015.

Page 286: Obama stresses "moral obligation" to fight climate change: White House, "Fact Sheet: President Obama to Announce Historic Carbon Pollution Standards for Power Plants," 3 August 2015.

Page 286: Pope's encyclical: Pope Francis, *Laudato Si*, 24 May 2015. Other faith leaders call for climate action: Bartholomew (leader of the Eastern Orthodox Church) and Justin Welby (archbishop of Canterbury), "Climate Change and Moral Responsibility," *New York Times*, 19 June 2015; Mitch Hescox, Evangelical Environmental Network, "Statement on Papal Encyclical," 18 June 2015; Arthur Neslen, "Islamic Leaders Issue Bold Call for Rapid Phase Out of Fossil Fuels," *Guardian*, 18 August 2015. Americans see climate change in moral terms: Ipsos/Reuters Poll, "Climate Change," 26 February 2015.

Page 287: Peabody campaign on global energy poverty: Peabody Energy, "Advanced Energy for Life Campaign Launched to Build Awareness and Support to End 'World's Number One Human and Environmental Crisis' of Global Energy Poverty," news release, 26 February 2014. "The only way that the folks in India are going to get a light bulb": Bob Murray, "Murray: My Goal to Be the Last Man Standing!" *Coal News*, July 2014, 31.

Page 287: India coal rush: Lisa Friedman and ClimateWire, "India Has Big Plans for Burning Coal," *Scientific American*, 17 September 2017; Gardiner Harris, "Coal Rush in India Could Tip Balance on Climate Change," *New York Times*, 17 November 2014. Pollution in New Delhi: Gardiner Harris, "Cities in India Among the Most Polluted, W.H.O.

Says," *New York Times*, 8 May 2014. 100,000 Indians die yearly: Lisa Friedman, "Coal-Fired Power in India May Cause More Than 100,000 Premature Deaths Annually," *Scientific American*, 11 March 2013.

Page 287: Indian anti-coal protests sometimes violently suppressed: Ted Nace, "Down with Coal! The Grassroots Anti-coal Movement Goes Global," *Grist*, 28 May 2011; CoalSwarm and Center for Media and Democracy, "Opposition to Coal in India," *SourceWatch*, accessed 28 April 2015. Coal rushes in other developing nations: Jan Cristophe Steckel, Ottmar Edenhofer, and Michael Jakob, "Drivers for the Renaissance of Coal," *Proceedings of the National Academy of Sciences*, vol. 112 (July 21, 2015); IEA, *Southeast Asia Energy Outlook*, 2013.

Page 288: Renewables can often meet needs more cheaply than coal: Carbon Tracker Initiative, *Energy Access: Why Coal Is Not the Way Out of Energy Poverty*, November 2014; Simon Bradshaw, *Powering Up Against Poverty: Why Renewable Energy Is the Future*, Oxfam Australia, 2015.

Bibliography

Adams, Charles Francis, Jr. *Railroads: Their Origin and Problems.* 1878. Reprint, New York: Arno Press, 1981.

Alinsky, Saul D. *John L. Lewis: An Unauthorized Biography.* 1949. Reprint, New York: Vintage Books, 1970.

Andersson, J. Gunnar. *Children of the Yellow Earth: Studies in Prehistoric China.* New York: Macmillan, 1934.

Anonymous. *A Collection of Very Valuable and Scarce Pieces Relating to the Last Plague in the Year 1665.* London: F. Roberts, 1721.

Ashton, T. S., and Joseph Sykes. *The Coal Industry of the Eighteenth Century.* 2d. ed. Manchester: Manchester University Press, 1964. First edition published in 1929.

Attenborough, David. *Life on Earth: A Natural History.* Boston: Little, Brown and Co., 1979.

Aurand, Harold W. *From the Molly Maguires to the United Mine Workers: The Social Ecology of an Industrial Union, 1869–1897.* Philadelphia: Temple University Press, 1971.

Baldwin, Leland D. *Pittsburgh: The Story of a City, 1750–1865.* Pittsburgh: University of Pittsburgh Press, 1937.

Becker, Jasper. *Hungry Ghosts: Mao's Secret Famine.* New York: An Owl Book, Henry Holt and Co., 1998.

Beecher, Catherine E., and Harriet Beecher Stowe. *The American Woman's Home: or, Principles of Domestic Science; Being A Guide to the Formation and Maintenance of Economical, Healthful, Beautiful, and Christian Homes.* New York: J. B. Ford and Co., 1869.

Bodde, Derk. *China's First Unifier: A Study of the Ch'in Dynasty As Seen in the Life of Li Ssu.* Hong Kong: Hong Kong University Press, 1967.

Bogen, Jules I. *The Anthracite Railroads: A Study in American Railroad Enterprise.* New York: Ronald Press, 1927.

Briggs, Asa. *Victorian Cities.* New York: Harper and Row, 1963.

_____. *The Power of Steam: An Illustrated History of the World's Steam Age.* Chicago: University of Chicago Press, 1982.

Brimblecombe, Peter. *The Big Smoke: A History of Air Pollution in London Since Medieval Times.* London and New York: Methuen, 1987.

Brown, Dee. *The Year of the Century: 1876.* New York: Charles Scribner's Sons, 1966.

Brown, Lester, Christopher Flavin, and Hilary French. *State of the World: 1999: A Worldwatch Institute Report on Progress Toward a Sustainable Society.* New York: W. W. Norton, 1999.

Bruere, Robert W. *The Coming of Coal.* New York: Association Press, 1922.

Bryner, Gary C. *Blue Skies, Green Politics: The Clean Air Act of 1990 and Its Implementation.* Washington, D.C.: CQ Press, 1995.

Burner, David. *Herbert Hoover: A Public Life.* New York: Alfred A. Knopf, 1979.

Carlson, Ellsworth C. *The Kaiping Mines, 1877–1912.*, 2d. ed. Harvard East Asian Monographs, Cambridge, Mass.: Harvard University Press, 1971.

Carpenter, John R., and Philip M. Astwood. *Plate Tectonics for Introductory Geology.* Dubuque, Iowa: Kendall/Hunt, 1983.

Carr, Donald E. *The Breath of Life.* London: Victor Gollancz Ltd., 1965.

Carter, Charles Frederick. *When Railways Were New.* New York: Simmons-Boardman, 1926.

Chadwick, Edwin. *Report on the Sanitary Conditions of the Labouring Population of Great Britain.* 1842. Reprint, edited and with an introduction by M. W. Flinn. Edinburgh: Edinburgh University Press, 1965.

Chang, Jung. *Wild Swans: Three Daughters of China.* New York: Anchor Books, 1992.

Chen, Cheng-siang. *China: Essays in Geography.* Hong Kong: Joint Publishing Co., 1984.

Church, R. A., ed. *The Coal and Iron Industries.* Cambridge, Mass.: Basil Blackwell, Ltd., 1994.

Cleal, Christopher J., and Barry A. Thomas. *Plant Fossils: The History of Land Vegetation.* New York: Boydell Press, 1999.

Coal Industry Advisory Board, International Energy Agency. *Coal in the Energy Supply of China; Report of the CIAB Asia Committee.* Paris: Organisation for Economic Co-Operation and Development, 1999. Available at http://www.iea.org.

Coleman, McAlister. *Men and Coal.* New York: Farrar and Rinehart, 1943.

Cressey, George. *Land of the 500 Million: A Geography of China.* New York: McGraw Hill, 1955.

Davis, Ralph. *The Rise of the English Shipping Industry in the Seventeenth and Eighteenth Centuries.* New York: St. Martin's Press, 1962.

Debeir, Jean-Claude, Jean-Paul Deleage, and Daniel Hemery. *In the Servitude of Power: Energy and Civilization Through the Ages.* London: Zed Books, 1991.

Degler, Carl N. *The Age of the Economic Revolution, 1876–1900.* 2d ed. Glenview, Ill.: Scott, Foresman, 1977.

Delaware and Hudson Company. *A Century of Progress: History of the Delaware and Hudson Company, 1823–1923.* Albany: J. B.Lyon Co., 1925.

Dickerman, William Carter. *Steam and the Railroads.* Princeton: Princeton University Press, 1936.

Dorian, James P. *Minerals, Energy, and Economic Development in China.* Oxford: Clarendon Press, 1994.

Dubofsky, Melvyn, and Warren Van Tine. *John L. Lewis: A Biography.* Abridged ed. Urbana, Ill.: University of Illinois Press, 1986.

Eavenson, Howard N. *Coal Through the Ages.* 2d. ed. New York: American Institute of Mining and Metallurgical Engineers, 1939.

Ellis, Hamilton. *British Railway History: An Outline from the Accession of William IV to the Nationalisation of Railways, 1830–1876.* London: George Allen and Unwin Ltd., 1954.

Emerson, Ralph Waldo. *The Conduct of Life.* New and rev. ed. Boston: Houghton, Mifflin and Co., 1893. Originally published in 1860.

Engels, Friedrich. *The Condition of the Working Class in England.* 1844. Reprint, edited and with an introduction and notes by David McLellan. Oxford and New York: Oxford University Press, 1999.

Evelyn, John. *Fumifugium: or, The Inconvenience of the Aer, and Smoke of London Dissipated.* 1661. Reprint, London: B. White, 1772.

Fairbank, John King and Merle Goldman. *China: A New History, Enlarged Edition.* Cambridge, Mass., and London: The Belknap Press of Harvard University Press, 1999.

Ferguson, Margaret. *A Study of Social and Economic Factors in the Causation of Rickets.* London: H. M. Stationery Office, 1918.

Finley, Joseph E. *The Corrupt Kingdom: The Rise and Fall of the United Mine Workers.* New York: Simon and Schuster, 1972.

Fleming, G. T. *History of Pittsburgh and Environs.* New York: American Historical Society, Inc., 1922.

Flinn, Michael W. *1700–1830: The Industrial Revolution.* Vol. 2 of *The History of the British Coal Industry.* Oxford: Clarendon Press, 1984.

Freeman, Michael. *Railways and the Victorian Imagination.* New Haven: Yale University Press, 1999.

Galloway, Robert. *Annals of Coal Mining and the Coal Trade.* Vol. 1. 1898. Reprint, with an introduction by Baron F. Duckham, Newton Abbot, Devon: David and Charles, 1971.

———. *A History of Coal Mining in Great Britain.* 1882. Reprint, with an introduction by Baron F. Duckham, Newton Abbot, Devon: David and Charles, 1969.

Gao, Mobo C. F. *Gao Village: A Portrait of Rural Life in Modern China.* Honolulu: University of Hawaii Press, 1999.

Gaskell, P. *The Manufacturing Population of England, Its Moral, Social, and Physical Conditions, and the Changes Which Have Arisen from the Use of Steam Machinery; with an Examination of Infant Labour.* London: Baldwin and Cradock, 1833.

Gelbspan, Ross. *The Heat Is On: The High Stakes Battle Over Earth's Threatened Climate.* Reading, Mass: Addison-Wesley, 1997.

Gibson, Rowland R. *Forces Mining and Undermining China.* London: Andrew Melrose, Ltd., 1914.

Gillin, Donald G. *Warlord: Yen Hsi-Shan in Shansi Province, 1911–1949.* Princeton: Princeton University Press, 1967.

Glorieux, Francis H., ed. *Rickets.* New York: Raven Press, 1991.

Gluck, Elsie. *John Mitchell: Miner: Labor's Bargain with the Gilded Age.* New York: John Day Co., 1929.

Goudsblom, Johan. *Fire and Civilization.* New York: Penguin Books, 1992.

Granet, Marcel. *Chinese Civilization.* New York: Alfred A. Knopf, 1930.

Graunt, John. *Natural and Political Observations Made Upon the Bills of Mortality.* 1676. Reprint, edited by Walter F. Wilcox, Baltimore: Johns Hopkins Press, 1939.

Greene, Homer. *Coal and the Coal Mines.* Boston: Houghton, Mifflin and Co., 1894.

Gribbin, John, and Mary Gribbin. *Children of the Ice: Climate and Human Origins.* Cambridge, Mass., and Oxford: Basil Blackwell, 1990.

Hammond, J. L., and Barbara Hammond. *The Town Labourer, 1760–1832.* Vols. 1 and 2. 1917. Reprint, London: Guild Books, 1949.

Hatcher, John. *Before 1700: Towards the Age of Coal.* Vol. 1 of *The History of the British Coal Industry.* Oxford: Clarendon Press, 1993.

Hertsgaard, Mark. *Earth Odyssey: Around the World in Search of Our Environmental Future.* New York: Broadway Books, 1998.

Hindle, Brooke, ed. *Material Culture of the Wooden Age.* Tarrytown, N.Y.: Sleepy Hollow Press, 1981.

Hindle, Brooke, and Steven Lubar. *Engines of Change: The American Industrial Revolution, 1790–1860*. Washington, D.C.: Smithsonian Institution Press, 1986.

Holland, John. *The History and Description of Fossil Fuels, the Collieries, and Coal Trade of Great Britain*. London: Whittaker and Co., 1835.

Holt, Emily. *Encyclopaedia of Household Economy*. New York: McClure, Phillips, and Co., 1903.

Howe, G. Melvyn. *Man, Environment and Disease in Britain: A Medical Geography of Britain Through the Ages*. New York: Barnes and Noble Books, 1972.

Houghton, John. *Global Warming: The Complete Briefing*. Elgin, Ill.: Lion Publishing, 1994.

Hudson, Pat. *The Industrial Revolution*. London and New York: Edward Arnold, 1992.

Intergovernmental Panel on Climate Change (IPCC). *Climate Change 2001: The Scientific Basis*. Cambridge: Cambridge University Press, 2001. Available at http://www.ipcc.ch.

———. *Climate Change 2001: Impacts, Adaptation, and Vulnerability* Cambridge: Cambridge University Press, 2001. Available at http://www.ipcc.ch.

———. *Climate Change 2001: Mitigation*. Cambridge: Cambridge University Press, 2001. Available at http://www.ipcc.ch.

———. *Climate Change 2007: The Physical Science Basis*. Cambridge: Cambridge University Press, 2007.

———. *Climate Change 2013: The Physical Science Basis*. Cambridge: Cambridge University Press, 2013.

———. *Climate Change 2014: Mitigation of Climate Change*. Cambridge: Cambridge University Press, 2014.

———. *IPCC Special Report on Carbon Dioxide Capture and Storage*. Cambridge: Cambridge University Press, 2005.

Jakle, John A. *Images of the Ohio Valley: A Historical Geography of Travel, 1740 to 1860*. New York: Oxford University Press, 1977.

Jeffrey, E. C. *Coal and Civilization*. New York: Macmillan Co., 1925.

Jevons, H. Stanley. *The British Coal Trade*. New York: E.P. Dutton & Co., 1915.

Johnson, James P. *The Politics of Soft Coal: The Bituminous Industry from World War I Through the New Deal*. Urbana, Ill.: University of Illinois Press, 1979.

Johnson, Paul. *A History of the American People*. New York: Harper Collins, 1999.

Karnow, Stanley. *Mao and China: From Revolution to Revolution*. New York: Viking Press, 1972.

Kemble, Frances Ann. *Record of a Girlhood*. 2d ed. Vol. 2. London: Richard Bentley and Son, 1879.

Kinne, Helen, and Anna Cooley. *Foods and Household Management: A Textbook of the Household Arts*. New York: Macmillan Co., 1918.

Knowlton, Frank Hall. *Plants of the Past: A Popular Account of Fossil Plants*. Princeton: Princeton University Press, 1927.

Kolbert, Elizabeth. *The Sixth Extinction: An Unnatural History*. New York: Henry Holt and Co., 2014.

Korson, George. *Black Rock: Mining Folklore of the Pennsylvania Dutch*. Baltimore: Johns Hopkins Press, 1960.

Landes, David. *The Unbound Prometheus: Technological Change and Industrial Development in Western Europe from 1750 to the Present*. London: Cambridge University Press, 1969.

Laslett, John H. M., ed. *The United Mine Workers of America: A Model of Industrial Solidarity?* University Park, Pa: Pennsylvania State University Press, 1996.

Levine, David, and Keith Wrightson. *The Making of an Industrial Society: Whickham, 1560–1765*. Oxford: Clarendon Press, 1991.

Levy, Elizabeth, and Tad Richards. *Struggle and Lose, Struggle and Win: The United Mine Workers*. New York: Four Winds Press, 1977.

Li, Wenyan, and Lu Dadao, eds., *Industrial Geography of China*. Beijing and New York: Science Press, 1995.

Long, Priscilla. *Where the Sun Never Shines: A History of America's Bloody Coal Industry*. New York: Paragon House, 1989.

Lorant, Stefan. *Pittsburgh: The Story of an American City*. Garden City, N.Y.: Doubleday and Co., 1964.

Lu, Yingzhong. *Fueling One Billion: An Insider's Story of Chinese Energy Policy Development*. Washington D.C.: Washington Institute Press, 1993.

Macfarquhar, Roderick. *The Coming of the Cataclysm, 1961–1966*. Vol. 3 of *The Origins of the Cultural Revolution*. New York: Columbia University Press, 1997.

———. *The Great Leap Forward, 1958–1960*. Vol. 2 of *The Origins of the Cultural Revolution*. New York: Columbia University Press, 1983.

Marcus, Steven. *Engels, Manchester, and the Working Class*. New York: Random House, 1974.

McPhee, John. *Annals of the Former World*. New York: Farrar, Straus and Giroux, 1998.

Melosi, Martin V. ed. *Pollution and Reform in American Cities, 1870–1930*. Austin, Tex.: University of Texas Press, 1980.

Merson, John. *The Genius That Was China: East and West in the Making of the Modern World*. Woodstock, N.Y.: Overlook Press, 1990.

Miller, Donald L., and Richard E. Sharpless. *The Kingdom of Coal: Work, Enterprise and Ethnic Communities in the Mine Fields.* Philadelphia: University of Pennsylvania Press, 1985.

Morgan, Ted. *Wilderness At Dawn: The Settling of the North American Continent.* New York: Simon and Schuster, 1993.

Morin, G. A., et al. *Long-Term Historical Changes in the Forest Resource.* New York: United Nations, 1996.

Mumford, Lewis. *Technics and Civilization.* New York: Harcourt, Brace and Co., 1934.

Nash, George H. *The Life of Herbert Hoover: The Engineer, 1874–1914.* New York: W. W. Norton, 1983.

National Assessment Synthesis Team. *Climate Change Impacts on the United States: The Potential Consequences of Climate Variability and Change: Foundation.* New York: Cambridge University Press, 2001. Entire report available at http://www.usgcrp.gov.

_____. *Climate Change Impacts on the United States: The Potential Consequences of Climate Variability and Change: Overview.* New York: Cambridge University Press, 2000. Entire report available online at http://www.usgcrp.gov.

Needham, Joseph. *The Development of Iron and Steel Technology in China.* London: The Newcomen Society for the Study of the History of Engineering and Technology, 1958.

_____. *Physics and Physical Technology, Part III: Civil Engineering and Nautics.* Vol. 4 of *Science and Civilisation in China.* Cambridge: Cambridge University Press, 1971.

Needham, Joseph, and Peter J. Golas. *Chemistry and Chemical Technology, Part XIII: Mining.* Vol. 5 of *Science and Civilisation in China.* Cambridge: Cambridge University Press, 1999.

Nef, J. U. *The Rise of the British Coal Industry.* Vols. 1 and 2. 1933. Reprint, London: Frank Cass and Co., 1966.

Nicolls, William Jasper. *Coal Catechism.* Philadelphia: J. B. Lippincott Co., 1915.

_____. *The Story of American Coals.* Philadelphia: J. B. Lippincott Co., 1904.

Nourse, Timothy. *Campania Foelix.* 1700. Reprint, New York and London: Garland Publishing, 1982.

Noyes, Robert, ed. *Handbook of Pollution Control Processes.* Park Ridge, N.J.: Noyes Publications, 1991.

Painter, Nell Irvin. *Standing At Armageddon: The United States, 1877–1919.* New York: W. W. Norton, 1987.

Park, Chris C. *Acid Rain, Rhetoric and Reality.* London and New York: Methuen, 1987.

Patterson, Thomas C. *Archaeology: The Historical Development of Civilizations.* Englewood Cliffs, N. J.: Prentice Hall, 1993.

Perlin, John. *A Forest Journey: The Role of Wood in the Development of Civilization.* New York: W. W.Norton, 1989.

Phelan, Craig. *Divided Loyalties: The Public and Private Life of Labor Leader John Mitchell.* Albany: State University of New York Press: 1994.

Pohs, Henry A. *The Miner's Flame Light Book: The Story of Man's Development of Underground Light.* Denver: Flame Publishing Co., 1995.

Polo, Marco. *The Travels of Marco Polo, the Venetian.* Edited by M. Komroff. Garden City, N.Y.: Garden City Pub. Co., 1930.

Preston, Diana. *The Boxer Rebellion: The Dramatic Story of China's War on Foreigners That Shook the World in the Summer of 1900.* New York: Walker & Co., 2000.

Ransom, P.J.G. *The Victorian Railway and How It Evolved.* London: Wm. Heinemann Ltd., 1990.

Reader, John. *The Rise of Life: The First 3.5 Billion Years.* New York: Knopf, 1986.

Richardson, Richard. *Memoir of Josiah White.* Philadelphia: J. B. Lippincott Co., 1873.

Robertson, Una A. *The Illustrated History of the Housewife, 1650–1950.* New York: St. Martin's Press, 1997.

Rolt, L.T.C. *George and Robert Stephenson, The Railway Revolution.* London: Longmans, Green and Co., 1960.

Rolt, L.T.C., and J. S. Allen. *The Steam Engine of Thomas Newcomen.* New York: Science History Publications, 1977.

Rule, John. *The Labouring Classes in Early Industrial England, 1750–1850.* London and New York: Longman, 1986.

Schlegel, Marvin W. *Ruler of the Reading: The Life of Franklin B. Gowen, 1836–1886.* Harrisburg, Pa.: Archives Publishing Company of Pennsylvania, 1947.

Schofield, Robert E. *The Lunar Society of Birmingham: A Social History of Provincial Science and Industry in Eighteenth-Century England.* London: Oxford University Press, 1963.

Schulman, Daniel. *Sons of Wichita: How the Koch Brothers Became America's Most Powerful and Private Dynasty.* New York: Grand Central Publishing, 2014.

Schurr, Sam. H., et al. *Energy in the American Economy, 1850–1975: An Economic Study of Its History and Prospects.* Baltimore: Johns Hopkins Press, 1960.

Seltzer, Curtis. *Fire in the Hole: Miners and Managers in the American Coal Industry.* Lexington: University Press of Kentucky, 1985.

Sharpe, J. A. *Early Modern England: A Social History, 1550–1760.* London and Baltimore: Edward Arnold, 1987.

Shurick, A. T. *The Coal Industry.* Boston: Little, Brown, and Co., 1924.

Simmons, Jack. *The Victorian Railway.* New York: Thames and Hudson, 1991.

Smil, Vaclav. *China's Environmental Crisis: An Inquiry Into the Limits of National Development*. Amonk, N.Y., and London: M. E. Sharpe, 1993.

_____. *Energy in World History*. Boulder, Colo.: Westview Press, 1994.

Smiles, Samuel. *The Life of George Stephenson, Railway Engineer*. Boston: Ticknor and Fields, 1859.

_____. *Lives of Boulton and Watt*. London: John Murray, 1865.

Smith, Alan G. R. *The Emergence of a Nation State: The Commonwealth of England, 1529–1660*. New York: Longman, 1984.

Stearns, Peter N. *The Industrial Revolution in World History*. 2d ed. Boulder, Colo.: Westview Press, 1998.

Stewart, Wilson N. *Paleobotany and the Evolution of Land Plants*. Cambridge and New York: Cambridge University Press, 1983.

Stradling, David. *Smokestacks and Progressives: Environmentalists, Engineers, and Air Quality in America, 1881–1951*. Baltimore: Johns Hopkins University Press, 1999.

Strasser, Susan. *Never Done: A History of American Housework*. New York: Pantheon Books, 1982.

Strouse, Jean. *Morgan: American Financier*. New York: Random House, 1999.

Sullivan, Walter. *Continents in Motion: The New Earth Debate*. New York: American Institute of Physics, 1991.

Sung, Ying-Hsing. *T'ien-Kung K'ai-Wu: Chinese Technology in the Seventeenth Century*. 1637. Reprint, translated by E-tu Zen Sun and Shiou-Chuan Sun, University Park, Pa.: Pennsylvania State University Press, 1966.

Taylor, Thomas N., and Edith L. Taylor. *The Biology and Evolution of Fossil Plants*. Englewood Cliffs, N.J.: Prentice Hall, 1993.

Taylor, W. Cooke. *Notes of a Tour in the Manufacturing Districts of Lancashire*. 1842. Reprint, with an introduction by W. H. Chaloner. 3d ed. London: Frank Cass and Co., 1968.

Teiwes, Frederick C., with Warren Sun. *China's Road to Disaster: Mao, Central Politicians, and Provincial Leaders in the Unfolding of the Great Leap Forward, 1955–1959*. Armonk, New York: M. E. Sharpe, 1999.

Thomas, Brinley. *The Industrial Revolution and the Atlantic Economy: Selected Essays*. London and New York: Routledge, 1993.

Thompson, E. P. *The Making of the English Working Class*. New York: Vintage Books, 1966.

Thurston, Anne F. *A Chinese Odyssey: The Life and Times of a Chinese Dissident*. New York: Charles Scribner's Sons, 1991.

Tocqueville, Alexis de. *Journeys to England and Ireland*. Edited by J. P. Mayer. London: Faber and Faber, 1963. Originally written in 1835.

Tregear, T. R. *An Economic Geography of China*. London: Butterworth and Co., 1970.

Trevelyan, G. M. *English Social History: A Survey of Six Centuries, Chaucer to Queen Victoria*. London and New York: Longmans, Green and Co., 1947.

Tuchman, Barbara. *A Distant Mirror: The Calamitous Fourteenth Century*. New York: Knopf, 1978.

Tuttle, Carolyn. *Hard at Work in Factories and Mines: The Economics of Child Labor During the British Industrial Revolution*. Boulder, Colo.: Westview Press, 1999.

Ure, Andrew. *Philosophy of Manufactures or An Exposition of the Scientific, Moral, and Commercial Economy of the Factory System of Great Britain*. 1835. Reprint, 3d ed., updated by P. L. Simmonds, New York: Burt Franklin, 1969.

Van Rensselaer, Martha. *A Manual of Home-Making*. New York: Macmillan Co., 1921.

Waggoner, Madeline Sadler. *The Long Haul West: The Great Canal Era, 1817–1850*. New York: G. P. Putnam's Sons, 1958.

Wallace, Anthony F.C. *St. Clair: A Nineteenth Century Coal Town's Experience with a Disaster-Prone Industry*. New York: Alfred A. Knopf, 1987.

Wang, Kung-ping. *Controlling Factors in The Future Development of the Chinese Coal Industry*. New York: King's Crown Press, 1947.

Ward, J. T. *The Factory System and Society*. Vol. 2 of *The Factory System*. Newton Abbot, Devon, U.K.: David and Charles, 1970.

Wellington, Duke of. *Wellington and His Friends: Letters of the First Duke of Wellington to the Rt. Hon. Charles and Mrs. Arbuthnot, the Earl and Countess of Wilton, Princess Lieven, and Miss Burdett-Coutts*. New York: St. Martin's Press, 1965.

Wendt, Herbert. *Before the Deluge: The Personalities, Discoveries, and Controversies of the Colorful Individuals Who Have Traced the Story of Life on Earth*. Translated by Richard and Clara Winston. Garden City, N.Y.: Doubleday, 1968.

Whitney, Gordon G. *From Coastal Wilderness to Fruited Plain: A History of Environmental Change in Temperate North America 1500 to the Present*. Cambridge: Cambridge University Press, 1994.

Williams, Frederick S. *Our Iron Roads: Their History, Construction, and Social Influences*. London: Ingram, Cooke, and Co., 1852.

Williamson, Alexander. *Journeys in North China, Manchuria and Eastern Mongolia, with Some Account of Corea*. London: Smith, Elder and Co., 1870.

Wilson, Richard, and Spengler, John, eds. *Particles in Our Air: Concentrations and Health Effects*. Cambridge, Mass.: Harvard University Press, 1996.

Wohl, Anthony S. *Endangered Lives: Public Health in Victorian Britain*. Cambridge, Mass.: Harvard University Press, 1983.

Woodward, John. *An Attempt Towards a Natural History of the Fossils of England.* London: F. Fayram, 1729.

_____. *An Essay Toward a Natural History of the Earth and Terrestrial Bodies, Especially Minerals.* London: Ric. Wilkin, 1695.

Wright, Tim. *Coal Mining in China's Economy and Society, 1895–1937.* Cambridge: Cambridge University Press, 1984.

Wrigley, E. A. *Continuity, Chance and Change: The Character of the Industrial Revolution in England.* Cambridge: Cambridge University Press, 1988.

Xu, Dixin, and Wu Chengming, eds. *Chinese Capitalism, 1522–1840.* Translated by Li Zhengde et al. New York: St. Martin's Press, 2000. First published in Chinese in 1985.

Yergin, Daniel. *The Prize: The Epic Quest for Oil, Money, and Power.* New York: Touchstone, 1993.

Young, Alexander. *Chronicles of the First Planters of the Colony of Massachusetts Bay, from 1623 to 1636.* Boston: Charles C. Little and James Brown, 1846.

Acknowledgments

Writing a book of this sort requires a sustained obsession over a single topic. I owe a great deal of thanks to all those who not only tolerated but indulged my obsession. Some people sent me news items relating to coal, some discussed the book with me as it evolved, some reviewed bits of the manuscript, and some provided crucial encouragement when I needed it. In particular, I want to thank Leslie Freese, Ginny Chase, Lynn James, Rich James, Sherry Coben, Paul Coben, Patricia Lawrence, John Whitmore, Lisa Tiegel, Gwen Macsai, Ross Gelbspan, Charles Medawar, Elizabeth McGeveran, Peter Ciborowski, and Sudie Hoffmann. I'm especially grateful to Betsy Sansby and Alan Dworsky for the many hours of creative input, for the enthusiastic support, and for inspiring by example.

Books like this also require access to an enormous library system. I'm grateful to the University of Minnesota, with its broad collection and helpful librarians, for providing it.

I offer a special thanks to the many people who generously assisted me in China, and particularly to Jane. They generally chose not to be identified in this book, in part because they tended to minimize the value of their assistance, but without them this book would have been considerably diminished.

I am hugely indebted to my agent, Robert Shepard. Without Robert's early and unflagging enthusiasm, his hard work, and his critical insights, this book might not exist. And my deep thanks go to Amanda Cook at Perseus, who took a chance in acquiring this book and then rose to the daunting challenge of editing it with grace and skill. Her ideas for shaping and streamlining the book have been invaluable. I am also grateful to Marietta Urban and Jennifer Blakebrough-Raeburn for helping turn the manuscript into a polished publication.

Finally, my deepest thanks go to my children, Tom and Ella, for their whole-hearted support and amazing patience, and to my incomparable husband, Jim Coben. Jim helped make this book a reality in more ways than I can count, and not least by agreeing that putting my legal career on hold to write a book about coal was an excellent idea and never wavering in that belief.

Illustration Credits

Carboniferous forest: From The Age of Reptiles, a mural by Rudolph F. Zallinger. Copyright © 1966, 1975, 1985, 1989, Peabody Museum of Natural History, Yale University, New Haven, CT.

Person dragging coal: From the Report of the Royal Commission, 1840.

Cotton mill: From *The Life and Adventures of Michael Armstrong, the Factory Boy*, by Frances Trollope (London: Henry Colburn, 1840).

Industrial housing: Thomas Annan, Photographs of the Old Closes and Streets of Glasgow, 1868/1877.

Child with rickets: From *The New Public Health: An Introduction for Midwives, Health Visitors, and Social Workers*, by Fred Grundy (London: H. K. Lewis, 1960).

Liverpool and Manchester Railway: Yale Center for British Art, Paul Mellon Collection.

Anthracite Canal: Pennsylvania Canal Society Collection, National

Canal Museum, Easton, PA.

Corliss Steam Engine: From *The Illustrated History of the Centennial Exhibition*, by James D. McCabe (Cincinatti: Jones Bros., and Co., 1876).

Breaker boys: National Archives, photo no. 102-LH-1938.

Women at stove: Wisconsin Historical Society, photo no. WHi (X3) 22443.

Pittsburgh pollution: Carnegie Library of Pittsburgh.

Cartoon: Chicago Record-Herald, 19 April 1909.

Other photographs by author.

Index

BARBARA FREESE is an environmental attorney and energy policy analyst, who for more than twelve years helped enforce her state's environmental laws as an assistant attorney general in Minnesota. Freese lives in St. Paul, Minnesota.